D0282429

Process Level
Instrumentation
and Control

ENGINEERING MEASUREMENTS AND INSTRUMENTATION

A Series of Reference Books and Textbooks

Editor

Paul N. Cheremisinoff

New Jersey Institute of Technology
Newark, New Jersey

Additional Volumes in Preparation

Process Level Instrumentation and Control

NICHOLAS P. CHEREMISINOFF

MARCEL DEKKER, INC.   *New York and Basel*

Library of Congress Cataloging in Publication Data

Cheremisinoff, Nicholas P.
  Process level instrumentation and control.

  (Engineering measurements and instrumentation;
v. 2)
  Includes bibliographical references and index.
  1. Process control.   2. Level indicators.   I. Title.
II. Series.
TS156.8.C538   629.8   80-21922
ISBN 0-8247-1212-9

COPYRIGHT © 1981 BY MARCEL DEKKER, INC. ALL RIGHTS RESERVED.

Neither this book nor any part may be reproduced or transmitted in any
form or by any means, electronic or mechanical, including photocopying,
microfilming, and recording, or by any information storage and retrieval
system, without permission in writing from the publisher.

MARCEL DEKKER, INC.
270 Madison Avenue, New York, New York 10016

Current printing (last digit):
10 9 8 7 6 5 4 3 2 1

PRINTED IN THE UNITED STATES OF AMERICA

Preface

Industrial systems range from simple processes to exceedingly complex operations and, within each phase of an operation, some type of mechanism or mechanisms must be applied in order to gauge process performance and correct system malfunctions. One of the most commonly encountered, and an often difficult, process variable to control is that of process medium level. Inaccurate detection of this parameter can lead to poor product quality, operation malfunctions, and system downtimes. Inaccuracies in inventory gauging can lead to misuse of storage volumes, which can result in many thousands of dollars in lost processing revenues. These problems become even more damaging when energy products (such as petroleum constituents) are involved.

Efficient utilization of process holdup volumes and various operations controls ensuring desired product qualities can only be achieved through proper process systems design and by the use of carefully selected and operated process instrumentation. The selection basis for any mode of parameter detection and control instrumentation depends heavily on a thorough understanding of the operation principles of the devices being considered. Such knowledge of the limitations, advantages, and past success of the instrumentation is data essential to proper selection of a system for a specific application.

In this volume are discussed the operating principles behind the major types of level detection instrumentation for both solid and liquid media handling. The limitations and operating experiences of each device are presented along with numerous examples of industrial applications. Design criteria for incorporating these systems into control schemes are also included.

This book was written with two audiences in mind—namely, practicing engineers and senior- to graduate-status engineering students.

It is hoped that the material presented herein will provide a working
guide for the industry segment in solving and providing ideas toward
the resolution of difficult level control problems. For students, this
volume is designed to provide exposure to the practical side of control
system design in the area of level control. All too often in the class-
room, the details of instrumentation design and application are not
given the attention warranted. It is assumed that the reader has had
an engineering and mathematical background and consequently the fundamental aspects of control theory are not stressed. A review of
basic control theory is, however, included throughout the book and in
an appendix.

Heartfelt gratitude goes to my many friends in industry who gave
of their valuable time, expertise, and materials, in order to make this
volume possible.

Nicholas P. Cheremisinoff

Contents

Process Level Instrumentation and Control

1

Principles of Level Measurement and Control

INTRODUCTION

Level measurement can be defined as the determination of the position of an interface existing between two mediums, separable by gravity, with respect to an established datum plane. Such interfaces can exist between a liquid and a gas, two liquids, a granular or fluidized solid and a gas, or a liquid and its vapor. There are numerous situations in industry where interfaces must be established within specified limits for purposes of process and product quality controls.

There are a variety of techniques available for measuring both liquid and solid levels in process equipment. The selection of instrumentation for level control and/or measurement can be confusing partly because many variations of these techniques are available commercially, each of these having certain limitations. Proper selection depends upon the nature of the process, allowable economics, and the degree of measurement accuracy and control required. It is essential that the user have a working knowledge of the various instruments available so that the appropriate selection and implementation can be made.

This book is intended as a guide for process engineers and scientists in selecting and using the best level measuring instruments and controls for their applications. It should be emphasized that the area of instrumentation is rapidly changing. New devices are constantly being introduced to the market; however, the working principles behind most new systems remains the same and generally only a more efficient device is developed. For this reason the fundamental theories behind the operation and design of some of the more sophisticated techniques have been stressed.

CLASSIFICATION OF INSTRUMENTS

Like many other process variables, level can be measured by direct or inferred methods. Direct methods make use of the following physical principles: bouyancy, fluid motion, density, and thermal, electrical, and optical properties. Inferential methods make use of hydrostatic head, displacement, and radioactive properties, as well as density.

The methods and instrumentation discussed in this book are classified as follows:

Visual techniques
    Dip stick
    Tape-and-plumb bob
    Open manometer
    Gauge glass

Float-actuated devices
    Chain or tape float gauge
    Lever-and-shaft mechanisms
    Magnetically coupled devices

Displacer devices
    Torque-tube displacer
    Magnetically coupled displacer
    Flexure-tube displacer

Head devices
    Pressure-gauge systems
    Bubble-tube systems
    Head systems

Electrical methods
    Electrical conductivity probes and devices
    Capacitance-type probes and devices

Thermal methods
    Temperature differential devices
    Thermal conductivity devices

Sonic methods
    Fixed-point detectors
    General sonic devices

Infrared devices

Nuclear devices

Table 1.1 can serve as an orientation chart and general selection guide for level detectors.

FUNDAMENTALS OF INSTRUMENTATION AND CONTROLS

The first step in analyzing any engineering problem is to prepare a drawing that details the flow of materials and/or heat to and from a system. Each part of the process can be represented by a block with input and output streams. Input and output variables are denoted by arrows to show the flow of information through the system.

Block diagrams are used in controls engineering to identify the instrumentation needed at various stages of each process. Instrumentation is represented on flow diagrams by a system of letters, numbers, and basic pictorial symbols. The symbols used to denote instrumentation are illustrated in Table 1.2. Instrumentation symbols used specifically for level are given in Table 1.3.

An understanding of the terminology of controls engineering is necessary to comprehend the fundamentals of level control. Considerable confusion has arisen among engineers because standardization of control terminology is largely incomplete; several terms often have been applied to the same control phenomenon. It is important, then, that the newcomer to this field be briefed on important definitions. The following terminology refers specifically to instrumentation and controls:

Automatic controller: A device which measures the value of a variable quantity and operates to correct or limit deviation of the measured value from a specified reference value. These devices provide both measuring and controlling services and contain one or more feedback loops, at least one of which comprises both the automatic controller and the process.

Table 1.1   Level Detector Orientation Chart and General Selection Guide (Y = yes, N = not applicable)

Gauging Characteristics	Hand-Line/Manual Gauges	Sight Gauges	Manometer	Float-Tape	Bubbler	DP Cell	Capacitance	Conductance	Resistance	Sonic/Sonar	Radio Frequency	Radioactive	Thermal	Photoelectric
No moving parts	N	Y	Y	N	Y	Y	Y	Y	Y	Y	Y	Y	Y	Y
Simple/rugged/reliable	Y	Y	N	N	N	Y	N	Y	Y	N	N	N	N	N
Can gauge viscous liquids/agitated slurries	Y	N	N	N	N	N	N	N	Y	Y	N	Y	N	Y
Can gauge fluent dry materials	Y	N	N	N	N	N	Y	Y	Y	Y	Y	Y	N	Y
Nonmetallic/chemically resistant	N	Y	Y	N		N	N	N	Y	Y	N	N	N	Y
Intrinsically safe group A, B, C, D, F, G	N	N	Y	N	N	N	Y	Y	Y	Y	Y	N	Y	Y
Measures level directly	Y	Y	N	Y	N	N	N	N	N	N	N	Y	N	N
Tank top accessible to install/service	Y	N	N	Y	Y	N	N	N	Y	Y	Y	Y	Y	N
No electronics at tank top	Y	Y	Y	Y	Y	N	N	N	Y	N	N	N	Y	N
Sensor/Gauge Unaffected by														
Material-specific gravity	Y	Y	N	N	N	N	N	N	Y	Y	N	Y	Y	N
Material temperature	Y	Y	N	Y	N	N	N	N	Y	Y-N	N	?	N	Y
Tank pressure/vacuum	Y	N	N	Y-N	N	N	N	N	Y	N	Y	Y	Y	Y
Tank/material humidity	Y	N	Y	N	Y	Y	N	N	Y	N	?	Y	N	Y
Foam/dust	Y	Y	Y	N	Y	Y	N	N	N	Y-N	?	?	Y	?

4

	1	2	3	4	5	6	7	8	9	10	11	12	13
Surface waves/agitation	N	N	Y	N	Y	Y	Y	Y	Y	N	N	N	?
Static electricity	Y	Y	Y	Y	Y	Y	Y	Y	Y	N	?	?	Y
Material Buildup/coating	N	N	N	N	N	N	N	N	N	Y	?	Y	N
Tank bottom contamination	N	N	N	N	N	N	N	Y-N	Y	Y	?	Y	Y
Material viscosity	Y	N	N	N	Y	Y	Y	Y	Y	Y	Y	Y	Y
Material conductivity	Y	Y	Y	Y	Y	Y	Y	Y	Y	N	Y	Y	Y
Offers combined level/temperature sensing	N	N	N	N	N	N	N	N	N	N	N	N	N
Fast response (wave gauging)	N	Y	Y	Y	Y	Y	Y	Y	Y	Y	Y	Y	Y
Applicable to short tanks	Y	Y	Y	Y	Y	Y	Y	Y	Y	Y	N	N	N
Calibration by precision simulation	N	N	N	N	N	N	N	Y	N	N	Y	Y	N
Level of technical skill required	Lo	Lo	Mod	Mod	Mod	Hi	Hi	Hi	Hi-Mod	Hi	Hi	Hi	Mod Hi

Nominal Accuracy

	1	2	3	4	5	6	7	8	9	10	11	12	13
H = 0.05% to 0.2%			Hi		Hi	Hi	Hi	Hi	Hi		Hi	Hi	Hi
Mod = 0.2% to 1%			Mod	Mod	Mod	Mod				Mod	Mod	Mod	Mod
Lo = over 1% error	Lo	Lo					Lo	Lo	Lo	Lo			

Relative Cost to

	1	2	3	4	5	6	7	8	9	10	11	12	13
Purchase	Lo	Lo	Mod	Mod	Mod	Mod	Mod	Mod	Mod	Mod	Hi	Mod	Hi
Install	Lo	Hi	Hi	Mod	Mod	Mod	Mod	Mod	Mod	Hi	Hi	Hi	Hi
Use/maintain	Lo	Lo	Mod	Hi	Hi	Mod	Hi	Hi	Mod-Hi	Hi	Hi	Hi	Hi

Table 1.2   General Instrumentation Symbols Used for Flow Plans

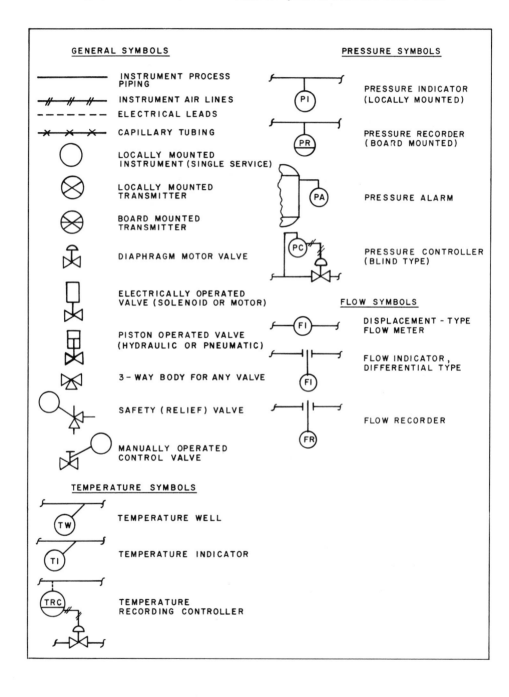

GENERAL SYMBOLS

INSTRUMENT PROCESS PIPING

INSTRUMENT AIR LINES

ELECTRICAL LEADS

CAPILLARY TUBING

LOCALLY MOUNTED INSTRUMENT (SINGLE SERVICE)

LOCALLY MOUNTED TRANSMITTER

BOARD MOUNTED TRANSMITTER

DIAPHRAGM MOTOR VALVE

ELECTRICALLY OPERATED VALVE (SOLENOID OR MOTOR)

PISTON OPERATED VALVE (HYDRAULIC OR PNEUMATIC)

3 - WAY BODY FOR ANY VALVE

SAFETY (RELIEF) VALVE

MANUALLY OPERATED CONTROL VALVE

TEMPERATURE SYMBOLS

TW   TEMPERATURE WELL

TI   TEMPERATURE INDICATOR

TRC   TEMPERATURE RECORDING CONTROLLER

PRESSURE SYMBOLS

PI   PRESSURE INDICATOR (LOCALLY MOUNTED)

PR   PRESSURE RECORDER (BOARD MOUNTED)

PA   PRESSURE ALARM

PC   PRESSURE CONTROLLER (BLIND TYPE)

FLOW SYMBOLS

FI   DISPLACEMENT - TYPE FLOW METER

FI   FLOW INDICATOR, DIFFERENTIAL TYPE

FR   FLOW RECORDER

Table 1.3   Level Instrumentation Symbols

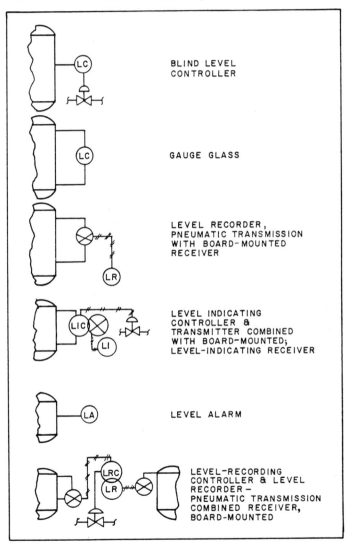

BLIND LEVEL
CONTROLLER

GAUGE GLASS

LEVEL RECORDER,
PNEUMATIC TRANSMISSION
WITH BOARD-MOUNTED
RECEIVER

LEVEL INDICATING
CONTROLLER &
TRANSMITTER COMBINED
WITH BOARD-MOUNTED;
LEVEL-INDICATING RECEIVER

LEVEL ALARM

LEVEL-RECORDING
CONTROLLER & LEVEL
RECORDER –
PNEUMATIC TRANSMISSION
COMBINED RECEIVER,
BOARD-MOUNTED

Automatic control system: This is an operable arrangement of one or more automatic controllers that are connected in closed loops with one or more processes.

Control point: The value of the controlled variable under any fixed set of conditions that the automatic controller is designed to maintain. Again, for those automatic controllers having two-position differential gap or floating with neutral control action, the control point is actually a range of values of the controlled variable.

Controlled variable: The quantity or condition which is measured and controlled.

Deviation: The difference between the actual value of the controlled variable and the controlled variable value that corresponds with the set point.

Error in measurement: The algebraic difference between a value resulting from measurement and the corresponding true value. When the error is positive, the measured value is algebraically greater than the true value.

Final control element: The portion of the controlling means which directly changes the value of the manipulated variable.

Manipulated variable: The quantity or condition which is varied by the automatic controller so as to affect the value of the controlled variable.

Measurement accuracy: The degree of correctness in which a measuring scheme produces the true value referred to by accepted engineering standards, such as the standard gram or meter.

Neutral zone: This represents a range of values of the controlled variable in which no change of position of the final control element takes place. It is usually expressed as a percentage of the controller scale range.

Offset: This is the steady-state difference between the control point and the controlled variable value that corresponds with the set point.

Precision: The degree of reproducibility among several independent measurements of the same true value under specified conditions.

Primary element: The portion of the measuring system which uses or transforms energy from the controlled medium to generate an effect which is a change in the value of the controlled variable.

Primary feedback: A signal which is related to the controlled variable and is compared with the reference input to obtain the actuating signal.

Relay-operated controller: A controller in which the energy transmitted through the primary element is either supplemented or amplified for operating the final control element by using energy from another source.

Resolution sensitivity: This represents the minimum change in the measured parameter which produces an effective response of the instrument. It is expressed either in units of the measured variable or as a fraction or percentage of the full-scale value.

Self-operated controller: This is a controller in which all the energy used to operate the final control element is derived from the controlled medium through the primary element.

Sensitivity: The ratio of output response to a change in the input. For measuring devices, the input is the measured variable, whereas for automatic controllers it is the controlled variable.

Set point: This represents the position at which the control-point setting mechanism is established. It denotes the position of the control-point setting mechanism on a controller translated into units of the controlled variable (the controlled variable in this case being level). Many types of automatic controllers are equipped with two-position differential gap, floating with neutral or proportional-position action. In this case, the set point is related to the position of a range of values of the controlled variable.

Servomechanism: This is an automatic control system where the controlled variable is in a mechanical position.

Threshold sensitivity: This is the lowest level of the measured variable which generates an effective response of the instrument or automatic controller.

MATHEMATICAL PRINCIPLES IN CONTROL THEORY

Instantaneous response to a change in variable represents an ideal
situation in control theory. For instance, when the level of a fluid
in a tank is controlled by an automatic controller, the response of
the control system may start immediately; however, a certain time
lapse will occur before the desired level is achieved. The time
element of any control system is dependent upon the responses of
the measuring means, the controlling means, and the process. As
such, the rate of change of level is equally as important as the magnitude of change.

    The time element is referred to as lag, which by general defini-
tion is the retardation of one physical condition with respect to
another physical condition to which it is related [1].

    The response of a thermometer bulb to temperature change is
one of the simplest systems illustrating the time lag of a control system. If a thermometer bulb is immersed in a bath of hot water, the
temperature change at the bulb is not detected instantaneously. Heat
must first be transmitted through the glass wall of the thermometer
to the filling medium. When the filling medium (assume mercury)
reaches a steady-state temperature a resultant change in fluid pressure, caused by thermal expansion of the filling medium, is transmitted to the receiving spiral. Figure 1.1 illustrates the temperature gradient for a mercury thermometer. The time lag of a thermometer bulb thus involves heat transmission as well as the dynamics of the moving element (spiral).

    The time response of any control system can be expressed by a
mathematical relationship called a transfer function. A transfer
function mathematically expresses the ratio of output signals to input signals of a system.

    The example of a mercury-filled, glass-bulb thermometer may
also be used to illustrate the general approach to developing such
mathematical relationships. A thermometer bulb is a first-order
system whose response can be described by a first-order linear
differential equation [2]. For first-order systems the energy—or
material—balance is

    Input - output = rate of accumulation   (1.1)

    From Equation (1.1), and assuming that the thermal resistance
of the glass medium is negligible in comparison to the film resistance
(i.e., it is assumed that the glass and the mercury would be at the

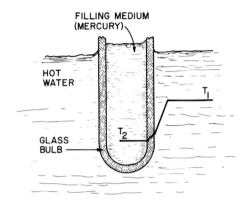

Figure 1.1   Illustrates the temperature gradients for a mercury thermometer.

same temperature), the unsteady-state heat balance can be written as follows:

$$UA\,(T_1 - T_2) - \theta = MC_p\frac{dT_2}{dt} = C'\frac{dT_2}{dt} \qquad (1.2)$$

where   M = mass of the mercury

C_p = heat capacity of the mercury

U = overall heat transfer coefficient

A = the heat transfer surface area

t = time

C' = thermal capacity of the system (i.e., the product of mass and specific heat)

The expression can be rearranged to solve for the input variable (T_1), as follows:

$$T_1 = \left(\frac{MC_p}{UA}\right)\frac{dT_2}{dt} + T_2 \qquad (1.3)$$

Note that the quantity MC_p/UA has units of time and as such is referred to as the time constant of the system. Equation (1.3) can thus be written as

$$T_1 = \theta \, \frac{dT_2}{dt} + T_2 \qquad (1.4)$$

where θ, the time constant, is a measure of the time needed by the system to adjust to a new input ($= MC_p/UA$). Note that Equation (1.4) states that the system's rate of response, dT_2/dt, is inversely proportional to the time constant. In a physical sense, the time constant for this example is the product of heat flow resistance ($1/UA$ or R) and storage capacity (MC_p or C'). In basic electrical theory the time constant is defined as the product of electrical resistance and electrical capacitance. Analogously, the time constant for fluid flow systems is the product of the capacity to store fluid and the flow resistance.

Equation (1.4) is a first-order linear differential equation that can be readily solved by means of Laplace transformations. Transforming a differential equation by Laplace transforms gives an algebraic equation, with a complex variable, called "s," that replaces time as the independent variable. Table 1.4 gives various Laplace transforms. For the reader who has forgotten how to use this mathematical tool, Appendix A provides a short review of the Laplace transform. There are also several problems at the end of this chapter that the reader may solve. (See also References 3-5 for detailed discussions on the use of Laplace transforms in control theory.)

The transfer function from Equation (1.4) is obtained by assuming that the initial value of T_2 is zero, so that the initial value term in the transformed expression drops out (this means that T_2 actually refers to deviations from the initial value). The Laplace transform is thus

$$\theta s T_2(s) + T_2(s) = T_1(s) \qquad (1.5)$$

(Note that s denotes the involvement of the transformed variables of T_1 and T_2).

Equation (1.5) can be rearranged to give the ratio of output to input of the system (i.e., the transfer function):

$$\psi \equiv \frac{T_2(s)}{T_1(s)} = \frac{1}{\theta s + 1} \qquad (1.6)$$

where ψ is the transfer function of the system.

The response of the thermometer to various input changes (i.e., changes in T_1) can now be determined by obtaining the inverse transform of Equation (1.6) after substituting in the transform of the particular input signal under consideration. For illustration, assume that the input is increased from zero to the value "a" at time $t = 0$. Then from Table 1.4

$$T_1(s) = \frac{a}{s} \qquad (1.7)$$

(Note in the untransformed state, $T_1 = a$).

and hence

$$T_2(s) = \frac{a}{s(\theta s + 1)} \qquad (1.8)$$

By taking the inverse transform of Equation (1.8), the actual temperature T_2 can be evaluated:

$$T_2 = \mathcal{L}^{-1}\left[\frac{a}{s(\theta s + 1)}\right]$$

$$= a\mathcal{L}^{-1}\left[\frac{a}{s(\theta s + 1)}\right] \qquad (1.9)$$

The solution is thus

$$T_2 = a\left(1 - e^{-t/\theta}\right) \qquad (1.10)$$

In a similar fashion, expressions for level control systems can be derived. The following discussion outlines the general approach by way of an example.

Table 1.4   Common Laplace Transforms

$f(s)$	$F(t)$
$\dfrac{1}{s}$	1
$\dfrac{1}{s^2}$	t
$\dfrac{1}{s^n}$ $(n = 1, 2, \cdots)$	$\dfrac{t^{n-1}}{(n-1)!}$
$\dfrac{1}{\sqrt{s}}$	$\dfrac{1}{\sqrt{\pi t}}$
$s^{-3/2}$	$2\sqrt{\dfrac{t}{\pi}}$
$\dfrac{1}{s-a}$	e^{at}
$\dfrac{1}{(s-a)^2}$	te^{at}
$\dfrac{1}{(s-a)^n}$ $(n = 1, 2, \cdots)$	$\dfrac{1}{(n-1)!} t^{n-1} e^{at}$
$\dfrac{1}{(s-a)(s-b)}$	$\dfrac{1}{a-b}(e^{at} - e^{bt})$
$\dfrac{s}{(s-a)(s-b)}$	$\dfrac{1}{a-b}(ae^{at} - be^{bt})$
$\dfrac{1}{(s-a)(s-b)(s-c)}$	$-\dfrac{(b-c)e^{at} + (c-a)e^{bt} + (a-b)e^{ct}}{(a-b)(b-c)(c-a)}$
$\dfrac{1}{s^2+a^2}$	$\dfrac{1}{a}\sin at$
$\dfrac{s}{s^2+a^2}$	$\cos at$

Table 1.4 Common Laplace Transforms (Cont)

$f(s)$	$F(t)$
$\dfrac{1}{s^2 - a^2}$	$\dfrac{1}{a}\sinh at$
$\dfrac{s}{s^2 - a^2}$	$\cosh at$
$\dfrac{1}{s(s^2 + a^2)}$	$\dfrac{1}{a^2}(1 - \cos at)$
$\dfrac{s}{(s^2 + a^2)^2}$	$\dfrac{t}{2a}\sin at$
$\dfrac{s^2}{(s^2 + a^2)^2}$	$\dfrac{1}{2a}(\sin at + at\cos at)$
$\dfrac{s}{(s^2 + a^2)(s^2 + b^2)}\ (a^2 \neq b^2)$	$\dfrac{\cos at - \cos bt}{b^2 - a^2}$
$\dfrac{1}{s^4 - a^4}$	$\dfrac{1}{2a^3}(\sinh at - \sin at)$
$\dfrac{s}{s^4 - a^4}$	$\dfrac{1}{2a^2}(\cosh at - \cos at)$
$\dfrac{s}{(s - a)^{3/2}}$	$\dfrac{1}{\sqrt{\pi t}}e^{at}(1 + 2at)$
$\sqrt{s - a} - \sqrt{s - b}$	$\dfrac{1}{2\sqrt{\pi t^3}}(e^{bt} - e^{at})$
$\dfrac{\sqrt{s}}{s - a^2}$	$\dfrac{1}{\sqrt{\pi t}} + ae^{a^2 t}\,\mathrm{erf}(a\sqrt{t})$
$\dfrac{1}{(s + a)\sqrt{s + b}}$	$\dfrac{1}{\sqrt{b - a}}e^{-at}\,\mathrm{erf}(\sqrt{b - a}\sqrt{t})$

MATHEMATICAL PRINCIPLES OF LEVEL CONTROL

The behavior of systems responding to level changes is often not as
straightforward as the example of the thermometer. Level control
systems can be placed into two broad categories: (1) those systems
where the level parameter is important as a process variable and
(2) those where the flow from the vessel is the important variable.
In the former, control systems are designed to maintain near constant levels in the presence of load changes. For the second group
of systems, the exact level is unimportant except at the extreme
operating ranges of the process where the vessel may run dry or
overflow. Surges in process flow conditions are offset by allowing
level to vary for the latter.

    For the purposes of our discussion at this point, let us consider
the system illustrated in Figure 1.2. As shown, the tank discharges
a fluid (say, water) through a control valve under conditions of atmospheric pressure. The pressure drop experienced by this system,
in feet of fluid, is equivalent to the liquid height in the vessel. Also
note that for water flowing through a constriction such as a valve, a
square-root law is followed:

$$m = \beta\sqrt{\Delta P} \tag{1.11}$$

where m = flow

           ΔP = frictional pressure loss

and β = parameter that is a function of valve-stem position.

For this example, assume β to be constant. Note that any pressure
drop in the discharge line can be included with the loss across the
valve, as the frictional pressure drop is proportional to the square
of the flow rate for turbulent conditions. Following the general form
of the material balance given by Equation (1.1), we can write

$$m_1 - \beta\sqrt{h} = A\left(\frac{dh}{dt}\right) \tag{1.12}$$

where h is the level height.

    Note that the solution to Equation (1.12) is straightforward only
for "step changes" in valve position or input flow. The expression
can, however, be approximated by a first-order linear differential

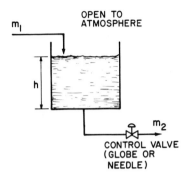

OPEN TO
ATMOSPHERE

m_1

h

m_2

CONTROL VALVE
(GLOBE OR
NEEDLE)

Figure 1.2 System under consideration. Liquid level is controlled by fluid flow rates into and out of tank.

equation by substituting a linear approximation for the effluent
term:

$$m_2 \simeq \overline{m} + \left(\frac{\overline{dm}}{dh} \right) (h - \overline{h}) \qquad (1.13)$$

This equation represents the tangent to the flow curve at the
average value of the flow, shown in Figure 1.3. The "bars" in
Equation (1.13) denote the average condition. In general, the linear
assumption is a good approximation for deviations of $\pm 20\%$ from the
average or mean value.

The transfer function can be further simplified by assuming the
variables flow and level to be mean deviations from the average
values in our equations. With this assumption the flow out of the
system can be simply written as

$$\overline{m}_2 = \overline{h} \frac{dm}{dh} \qquad (1.14)$$

The overall material balance becomes

$$\overline{m}_1 = A \frac{dh}{dt} + \overline{h} \frac{dm}{dh} \qquad (1.15)$$

or $\dfrac{\overline{m}_1}{dm/dh} = \boldsymbol{\theta}\,\dfrac{dh}{dt} + \overline{h}$ (1.16)

where A = cross-sectional area of the tank, and $\boldsymbol{\theta} = \dfrac{A}{dm/dh}$, the
time constant for the tank.

The resistance term in the time constant has the units of
driving force divided by the flow rate (dh/dm). The term resistance
can have several meanings in level control. For systems that are
strictly linear (i.e., flow being directly proportional to the fluid
driving force), resistance can be thought of as the total driving
force over the total flow rate. Such a system is illustrated in Figure 1.4, where resistance, R, is obtained from the inverse slope of
the flow curve. This definition, however, only applies to a strictly
linear system.

Resistance in general terms is defined as the rate of change of
driving force with flux or flow:

$$R = \frac{dh}{dm}$$ (1.17)

If an expression describing the flux is known, then the resistance
can be expressed in terms of the average flux and average driving
force. For the example presented above, let us assume that the
flow can be described by Equation (1.11). Then we can write

$$\frac{dm}{dh} = \frac{\beta}{2\sqrt{h}} = \frac{m}{2\sqrt{h}\sqrt{h}}$$ (1.18)

and an average value of the resistance becomes

$$\left(\frac{\overline{dh}}{dm}\right) = \frac{2\,\overline{h}}{\overline{m}}$$ (1.19)

Multiplying both sides of Equation (1.19) by A, an expression
for the time constant is obtained:

$$\boldsymbol{\theta} = \frac{2A\overline{h}}{\overline{m}} = \frac{2\overline{V}}{\overline{m}}$$ (1.20)

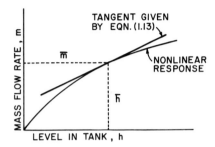

Figure 1.3   Illustrates flow curve for the system under consideration.

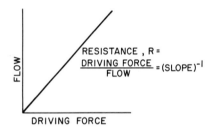

Figure 1.4   The dynamic resistance for a linear system is obtained from the inverse slope of the flow curve.

$\overline{V}$ represents the average volume of the fluid in the tank. Equation (1.20) states that the time constant is twice the residence time for the tank. The time constant derived is only applicable for the example under consideration; that is, for an open tank discharging to the atmosphere. Harriott [2] has derived the time constant for a tank under constant pressure [Equation (1.21)].

$$\theta = A\frac{dh}{dm} = 2A\frac{\overline{h} + h_0}{\overline{m}} \qquad (1.21)$$

The problems at the end of this chapter illustrate the usefulness of linearized approximations in describing level control systems.

NOMENCLATURE

A   area, lb/ft^2

a   constant in Equation (1.7), lb/ft^2

C_p specific heat of fluid Btu/(lb)(°F)

C'   thermal capacity (MC_p), Btu/°F

h   height of level, ft

M   mass, lb

m   flow rate, ft^3/h

P   pressure, lb/in^2

R   resistance [refer to Equation (1.17)], s/ft^2

s   denotes transformed variable

T   temperature, °F

t   time, s, h

U   overall heat transfer coefficient, Btu/(h)(ft^2)(° F)

V   volume, ft^3

Greek Letters

β valve stem position parameter [Equation (1.11)]

θ time constant, (s)

SUGGESTED STUDY PROBLEMS AND QUESTIONS

1.1   Write the Laplace transforms for the following equations and
       solve for x(s):

       a. $\frac{dx}{dt} + 3x = 5 \cos 2t$, where $x_0 = 1.5$

  b. $\dfrac{d^3x}{dt^3} + 2\dfrac{dx}{dt} + 2x + 3 = 0$, where x_0'', x_0', $x_0 = 0$

  c. $2\dfrac{d^2x}{dt^2} + \dfrac{dx}{dt} + 3 = A \tan 2\omega t$, where $x_0' = 1$, $x_0 = 2$

1.2 Using partial-fraction expansion and Laplace transforms, solve the following equation:

$$6\dfrac{dx}{dt} + x = 2t, \text{ where } x_0 = 0$$

1.3 A cylindrical tank has a diameter of 3 ft and an overall height of 9 ft. The level of liquid in the tank is normally at 5 ft. The flow to the tank is 12 ft^3/min ($\pm 10\%$). The fluid discharges through a control valve to a process that is under atmospheric pressure.

  a. Assuming the tank to be open to the atmosphere, compute the time constant.

  b. Assuming the tank is a closed system with a constant pressure of 15 psig above the liquid, compute the time constant.

1.4 An open cylindrical tank 5 ft in diameter is filled with a 10% caustic solution to a depth of 3.5 ft. The caustic solution flows through a 1-in. control valve at an average rate of 20 gal/min. Determine the new depth in the vessel if the input is increased by 30%.

1.5 For Problem 1.4 determine how long it will take to accomplish 95% of the change in level from the new flow condition.

1.6 For Problem 1.4, compute the transfer function relating tank level to valve position. Compare the results with the transfer function relating level and input flow.

1.7 A cylindrical tank has an area of 2.5 ft^2 and a normal liquid depth of 5 ft. The normal discharge rate is 187 gal/h. By use of a linearized solution, determine how the depth changes with time if the flow to the tank is suddenly increased by 30%.

1.8   For Problem 1.7, determine the exact solution. Plot the level height versus time for the two solutions and comment on the accuracy of the linearized approximation.

2

Design Practices in Level Control

INTRODUCTION

As noted in the previous chapter, the criteria for good level control
in process vessels differ greatly from those for control of other
process variables such as flow, pressure, or temperature. For example, the outlet temperature control of a furnace coil is established
on the basis of how much the temperature deviates from the setpoint
and on the speed of the temperature's return to that setpoint. Level
control in a process vessel supplying fuel to the furnace, however,
is evaluated on the basis of flow stability. For many applications
the actual value of level is not important within the range of the level
instrument. In the example of the furnace, the coil outlet temperature is established on the stability of the measured variable (i.e.,
temperature), whereas level control is rated on the basis of the
stability of the manipulated variable (flow). As such, the process
design basis for level control differs from the control of other variables in a process.

There are exceptions where the value of level does become important. One such example is the level in a settling basin. These
situations are described below. Also treated in this chapter are the
aspects of level measurement and control often considered in process design and operation. Basic principles applied in preparing
design specifications for level control systems are discussed.

PRINCIPLE TYPES OF LEVEL CONTROL ACTION

On-Off Control Systems

In general, on-off controllers are the least expensive and most
widely used type of control. Examples of their use include domestic
heating units, refrigeration systems, and water tanks. The basic
function of these systems is as follows. When the measured variable
falls below some set point, the controller comes on and the output
signal equals the maximum value. When the measured variable lies
above the set point, the controller is in the off position and the output signal is zero. It should be noted that because of mechanical
friction or arcing at electrical contacts, controllers actually over-
shoot the set point. This factor can sometimes be used to an advantage by deliberately increasing the interval or differential gap for
the purpose of decreasing the frequency of operation and reducing
wear.
      Simple on-off controllers have a major drawback in that they
often require too-frequent control valve operation or pump starts
and stops. For this reason, on-off with neutral-zone control action
is often preferred. For these systems, a control valve is fully
opened when the level approaches an upper set point which represents
the top of the neutral zone. The flow remains full until the level
drops to a lower set point (i.e., the bottom of the neutral zone),
where it is closed. Process applications where this type of control
is used include (1) applications where the flow rate is low and can
result in an easily plugged control valve, (2) when the flow is sent to
storage or disposal and flow stability is unimportant, (3) and when
liquids are discharged to a vessel in slugs.

Proportional Control Systems

Proportional control systems are among those most widely used
and are employed to achieve steady operation when disturbances are
absent and when the controlled variable must be a continuous function of the error [2]. The controller output with these systems is a
linear function of the error signal. The controller gain is defined
as the fractional change in output divided by the fractional change
in input:

$$\Delta p = K_c \, \Delta \epsilon \tag{2.1}$$

where Δp = fractional change in controller output,

   $\Delta \epsilon$ = fractional change in error, and

   K_c = controller gain.

   The control behavior of proportional controllers is often expressed by the proportional bandwidth. Bandwidth is the error
needed to cause a 100% change in controller output. The bandwidth
is expressed as a percentage of the chart width. As an example, a
bandwidth of 50% would mean that the controller output would go
from 0 to 1 for an error of 50% of the chart width. This can be
expressed as

$$b_w = \frac{1}{K_c} \times 100 \tag{2.2}$$

   For level control systems employing this scheme, the control
valve stem position is proportional to the level measurement. A
level control system based on proportional control is illustrated in
Figure 2.1. For the system shown, a level change equivalent to the
full range of the level instrument makes the valve stem move to its
full range. This, then, is an example of 100% proportional band
(i.e., the percentage figure is the amount of level change required to cause the valve stem to travel to its maximum position).

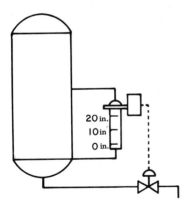

Figure 2.1   Example of a level control system using proportional
control.

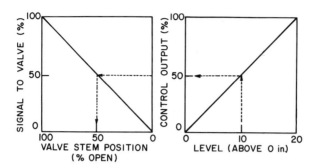

Figure 2.2   Illustrates the level-valve operation chart for 100% proportional band.

Figure 2.2 illustrates the level-valve operation chart for the 100% proportional band example.

It is of interest to note that we can force our system beyond the 100% proportional band. Figure 2.3 illustrates this situation. The plot shows for a proportional band greater than 100% that a full-range level change cannot result in full-range travel of the valve stem (i.e., the valve can neither be fully shut nor fully opened). As such, 100% represents the upper limit for proportional control action. If the proportional band is less than 100%, the reverse is true. In this case the full-range travel of the valve stem corresponds to only a portion of the level range (in terms of the control output versus level position plot, the curve shifts to the right and intercepts the x-axis). The low-proportional bands would then be employed for situations where maintaining the level close to the set point is important, but flow stability through the control valve is not.

Proportional control action has one other characteristic worth noting. That is, the steady-state level is strictly a function of the throughput and the valve size. As shown by the example in Figure 2.2, at a 10-in. level the valve is at the half-lift stem position. As the flow increases down the tower (Figure 2.1) to, say, a flow requiring 80% valve stem position, the level will have to stabilize at 16 in. This means that level control systems employing proportional control limited to 100% maximum proportional band are unable to return to a set point at changing flow rates.

A variation of proportional control allows level control systems to overcome this limitation. This is called integral control or

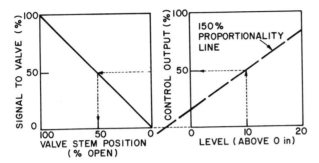

Figure 2.3   Illustrates the case where the proportional band is greater than 100%.

proportional-plus-reset control. The reset action readjusts the control signal to the set point by adding or subtracting increments to and from the control signal depending on the output signal's relation to the set point. The rate of change (i.e., rate of addition or subtraction) is proportional to the deviation of the level from the set point. Reset action thus continues to change the control valve signal until the level returns to the set point.

Reset control action with proportional control allows the proportional band to be adjusted above 100%; this further stabilizes the control valve flow rate and allows the level always to return to the set point value after an upset condition. These control systems are well suited for processes where liquid holdup is important and frequent upset conditions occur. They also allow maximum usage of the holdup built into a process vessel. A major drawback with proportional-plus-reset control action is that often a correction is made too fast, causing control instability problems.

Averaging Control Systems

Averaging control systems are another variation of proportional control. The term applies to proportional control plus reset action where the proportional band adjustment has been extended well beyond 100%. This causes the rate of flow through the control valve to change slowly. One of the problems in establishing wide proportional bands is the selection of the size of upset condition. When the proportional band is extended to allow full level change for small

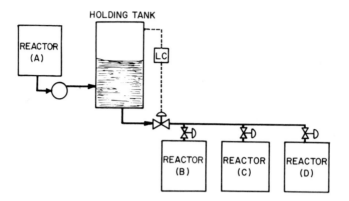

Figure 2.4   Illustrates an average level control system.

upsets, the large upsets will send the level beyond the range of the
instrument. In addition, the bands that are suitable for large upsets
may waste a portion of the holdup on smaller upsets.

      Averaging control is generally employed in cases where the
level itself is not important. An example of this situation is illustrated in Figure 2.4. In this example, the output stream from
reactor (A) is sent to a holding tank or storage vessel. A level controller (LC) regulates the flow to reactors (B), (C), and (D). When
one of the reactors in the downstream process is shut down, there
is a step change in flow to the storage vessel and the level starts to
drop. For such a system, a wide proportional band is employed on
the controller so that a substantial change in level takes place before
the outflow again equals the inflow condition. The change in outflow
is thus spread over a time period of several minutes so that disturbances to the downstream portion of the process are less severe
than would occur for a step change in flow. If a high-gain controller
is employed, the level would remain nearly constant and the effluent
flow rate would rapidly decrease to a new value. For this case the
storage capacity of the tank would be wasted with respect to flow
fluctuations.

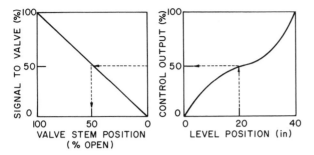

Figure 2.5   Illustrates the level-valve operational plot for the error-squared proportional control technique.

Error-Squared Control

These systems are employed to accommodate both large and small upsets in a process. They have the advantage of providing better usage of the holdup. Basically, these controllers employ a non-linear control action, referred to as error-squared proportional control. The relationship between the valve position and level is illustrated in Figure 2.5. The plots show that the magnitude of the valve position affected by a given change in level is least sensitive at the midrange. For these controller types the reset must always be added to error-squared control, or else the desirable flat portion of the curve would correspond with the actual valve position only when the valve is half open.

Inverse Derivative Control

Inverse derivative action is often added to conventional controllers to stabilize processes with noisy signals [6]. When a signal changes rapidly, the change in controller output is smaller than for a steady error of the same size [2, 6]. As such, the proportional band becomes widened for high-frequency signals but remains unchanged for low-frequency signals.

Levels that bounce up and down rapidly result from a variety of processes where boiling or splashing occur. Inverse derivative control action dampens the change in the control signal, so that the control valve employs an average position rather than following the unstable level. The net result is that the flow is more stable. In general, these are inexpensive systems and relatively small in size. They are often added in the air lines of a level control loop and can be added in the field as needed without being specified in the system design.

Cascade Control Systems

Cascade control is the best approach of using an additional controller for the purpose of decreasing upsets. In cascade control, the output of the primary controller is used in adjusting the set point of a secondary controller. The second controller then sends a signal to the control valve. Process output signals are fed back into the primary controller and signals from an intermediate segment of the process are relayed back to the secondary controller. In general, these systems provide the best performance for almost all types of load changes.

Harriot [2] notes that for disturbances introduced near the beginning of a system, the secondary controller begins corrective action before the process output indicates any deviation. The error may be on the orders of 10 to 100 times lower than that experienced with a single controller. Because cascade control has a higher natural frequency, the error integral may be only reduced two- to five-fold for distrubances introduced to the last stage of the process [7 - 9].

The secondary controller for level systems is usually a flow controller. The reason for positioning a flow controller between the primary controller and the control is to ensure that the flow does not deviate unless a change has been specified by the primary controller. This means that the flow controller counteracts any disturbances in its line that might otherwise cause a change in the flow rate.

Level/flow cascade control is primarily limited to the following general applications: (1) situations where numerous upsets and disturbances occur in lines containing the control valve, (2) when operators experience poor control with straight-level control schemes, and (3) when there is a need to reestablish questionable flow stability (e.g., this may take place on the second or third holdup in a series circuit).

Table 2.1   Illustrative Example of Level and Holdup Changes
for Step Changes in Input Flow Conditions

Normal Throughput (gal/min)	Step Change of 15% (gal/min)	Rate of Level Change (% of range/min)	Holdup Time (min)
2.1	700	105.0	7.0
4.3	350	52.5	3.5
15.0	100	15.0	1.0
30.0	50	7.5	0.5

LIQUID HOLDUP CONSIDERATIONS AND TANK DYNAMICS

Holdup or residence time becomes a critical parameter in many
operations. As shown by the expressions derived in Chapter 1, the
time constant of a control system is directly proportional to holdup.
There are a variety of situations that arise in process design where
it is necessary to determine the proper holdup needed in a vessel or
tower for level control application. The criterion for establishing
holdup speed of level changes for a given upset in flow conditions.
For a known change in flow rate at a given holdup volume, the rate
of level change is directly proportional to the throughput. As an
example, consider a vessel with a holdup of 1500 gal, assumed to be
within the level instrument's range. If a 15% step change in the normal
throughput were introduced to the system, the rate of level change
would occur in accordance with the tabulated values in Table 2.1.
The rate of throughput influences the ability of a given volume of
holdup to absorb variations in flow rate. For this reason holdup
is defined as the ratio of holdup volume to flow rate and is expressed
in dimensions of time. The last column in Table 2.1 reports holdup
values for the 1500-gal tank. In many applications, vessels are
sized in terms of minutes. This criterion allows ready conversion
to tank dimensions. It should be noted that in the context of level
control, only the holdup within the range of the instrument is of
prime importance.

The proper holdup time required to maintain flow stability in the control valve line of a process depends on a number of factors. The major parameters that should be considered in the design specifications of holdup for level control systems are

Range of flow specifications which are acceptable for the flow stability of the process: by knowing where the fluid goes (e.g., to tankage, distillation column, etc.), the necessary stability can be estimated.

Type and number of controller adjustments needed: often the specific controller adjustments will be an unknown and, as such, it is acceptable practice to assume that the adjustments will not deviate greatly from the optimum condition. Unfortunately this may be an unrealistic assumption for many cases.

Type, size, and frequency of upsets that may be encountered in a process: upset conditions are difficult to account for and usually are undefined at the time surge capacity is being established.

Because of the uncertainties and assumptions that must be made, holdup times are usually established based on specific experience with the type of process situation encountered. Holdup guidelines for two cases frequently encountered are given in Figure 2.6.

As noted throughout this chapter, the dynamics of level control are guided by the lags in the tank or vessel, the measuring device and the control valve. It was noted in Chapter 1 that for small step changes in level, a vessel with a control valve in the discharge line acts like a first-order system. The time constant for a tank is on the order of several minutes, and it should be noted that this is significantly greater than the other lags in the system.

When level controllers are used to regulate influent flow rate to a process vessel and the outflow is fixed (usually by a flow controller or a pump), then the tank can be characterized as a pure capacitance system and has no self-regulation. The level then lags the input by 90° at all frequencies. Note that large phase lags are not necessarily a disadvantage, since a tank with a time constant of several minutes would contribute nearly 90° lag at the critical frequency [2]. References 2, 3, and 10 should be consulted for discussions on critical frequency and maximum gains of control systems regulating input flows.

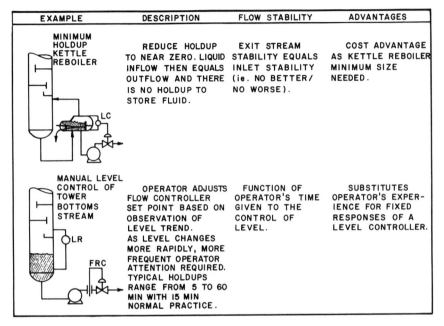

EXAMPLE	DESCRIPTION	FLOW STABILITY	ADVANTAGES
MINIMUM HOLDUP KETTLE REBOILER	REDUCE HOLDUP TO NEAR ZERO. LIQUID INFLOW THEN EQUALS OUTFLOW AND THERE IS NO HOLDUP TO STORE FLUID.	EXIT STREAM STABILITY EQUALS INLET STABILITY (ie. NO BETTER/ NO WORSE).	COST ADVANTAGE AS KETTLE REBOILER MINIMUM SIZE NEEDED.
MANUAL LEVEL CONTROL OF TOWER BOTTOMS STREAM	OPERATOR ADJUSTS FLOW CONTROLLER SET POINT BASED ON OBSERVATION OF LEVEL TREND. AS LEVEL CHANGES MORE RAPIDLY, MORE FREQUENT OPERATOR ATTENTION REQUIRED. TYPICAL HOLDUPS RANGE FROM 5 TO 60 MIN WITH 15 MIN NORMAL PRACTICE.	FUNCTION OF OPERATOR'S TIME GIVEN TO THE CONTROL OF LEVEL.	SUBSTITUTES OPERATOR'S EXPER- IENCE FOR FIXED RESPONSES OF A LEVEL CONTROLLER.

Figure 2.6   Holdup characteristics and guidelines for two examples.

SAFETY CONSIDERATIONS

A major consideration in all level control applications is safety.
Special precautions must be taken to ensure personnel safety in the
operation of level instrumentation and controls. Equipment selection requires special attention to possible explosion-proof requirements. In-depth study into tank explosions over the years has increased our knowledge of the potential dangers involved in handling
flammable liquids or vapors. For example, elimination of manual
dipping techniques for liquid storage gauging by automatic tank
gauging equipment has greatly increased personnel safety and installation protection. Automatic equipment must, however, meet very
specific safety requirements.
     Prior to specifying instrumentation, the potential degree of
hazard must be classified and the extent of the area in which the
hazard occurs defined. Flammable liquids are considered hazardous

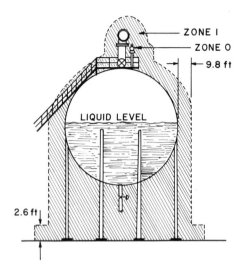

Figure 2.7   Hazardous area classification for a pressure sphere.
"Hazardous Classification Zones are designated in accordance with
the International Electrochemical Commission; see Reference 15."

if they have flash points below the maximum summer temperature,
or are heated to a temperature within 25°F of their flashpoints [11].
The reader should refer to the Associated Factory Mutual Fire Insurance Companies publications [12] for classifications on flammable
and combustible liquids.

The International Electrochemical Commission (IEC) has developed a means of hazard classification based on three zones for
a tank [13]:

Zone 0 represents an explosive gas-air mixture that is continuously present or present for an extended period of time.

Zone 1 represents an explosive gas-air mixture that is
likely to occur during normal operation.

Zone 2 represents an explosive gas-air mixture that is not
likely to occur and if it does, will only exist for a short
period of time.

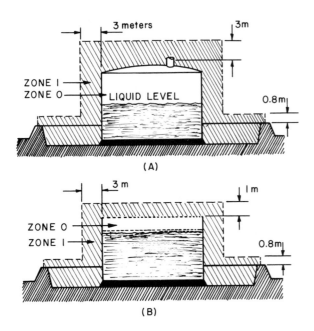

Figure 2.8 Hazardous area classifications for storage tanks with (A) fixed-roof installation and (B) floating-roof installation.

Figures 2.7 and 2.8 illustrate the hazard area classifications for different storage tanks as specified by the West German government [14, 15].

The IEC classifies five primary sources of hazards in vessels:

1. Open tanks or containers

2. Tanks without inert gas cushion or vents

3. Safety valves or any vents that release the tank atmosphere to the outside atmosphere

4. Pump glands, compressor glands, etc., which represent potential sources of leakage

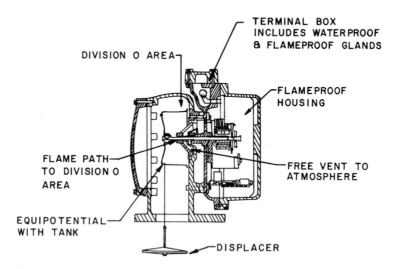

Figure 2.9   Illustrates one type of tank gage used for atmospheric tank storage service.

5.   Sample ports or valve outlets that release tank atmosphere to the working environment and are used frequently.

It should be noted that these classifications closely agree with the U.S. National Electrical Code [16].

Note that, in Figure 2.7 and 2.8, part of the level detection practices take place in Zone 0. The level detection equipment often forms the partition between Zone 0 and Zone 1. Because of this fact, special safety precautions must be included in the design and equipment layout. A variety of potentially dangerous situations can be eliminated by applying safety guidelines.

The primary elements that should be considered in selecting and installing level instrumentation include flame arresting, static charge buildup, sparking, electrical considerations, and lightning.

Flame-arresting effects should be examined for any vessel opening to the atmosphere. Any openings between Zone 0 and the atmosphere (refer to Figures 2.7 and 2.8) should be protected by an effective flame arrester. In this way a fire outside the area should not be able to reach Zone 0.

Figure 2.9 illustrates a tank gauge in which the shaft exiting Zone 0 of the tank is protected by a flame path having sufficient

length and width. The bushing shown in this example is constructed
of stainless steel.

The accumulation of static charges can lead to an explosive
condition. The region outside of Zone 0 should not be permitted to
accumulate an incendiary static charge. Reference 16 should be
consulted for the specific requirements for avoiding sparking hazards
caused by electrostatic charges. Standard practice is to ensure that
all nonconductive material inside Zone 0 be restricted within well-
defined limits or guidelines if the surface resistance is in excess of
10 Ω [10 - 16]. Plastic displacers on floats are generally not recommended; however, there are a few special circumstances. Metal
components inside the vessel must be at the tank potential and the
maximum permissible resistance should not exceed 10 Ω.

For the level gauge system shown in Figure 2.9, the drum compartment of the unit should be connected to the tank roof via a copper
strip. The stainless steel displacer, measuring wire, and drum
are all connected to the drum compartment by stainless steel bearings or carbon—PTFE bearings. Again, to prevent the accumulation
of static charges, the electrical resistance of the bearings should not
exceed 10 Ω. The presence of carbon in the bearings will reduce the
electrical resistance to within acceptable limits.

To prevent sparking, all fixed metal parts inside the vessel must
be sparkproof. The drum housing of the tank level gauge shown in
Figure 2.9 is made from aluminum and is immersed in an oil medium
that is impact-resistant. The thickness of this coating should not
exceed 0.2 mm in order to limit static charge buildup [11]. Further-
more, the maximum impact energy of components that can fall into
the tank should not exceed values below the safety margin of the
respective material.

To ensure electrical safety, no electrical equipment should be
used in Zone 0, with the exception of specifically certified circuits
that are intrinsically safe. It must be emphasized that flameproofed
electrical equipment will not provide acceptable safety in Zone 0.
When flameproof equipment is employed outside Zone 0, direct gas
contact between the flameproof enclosure and Zone 0 must be avoided.
Separation can be achieved by one of two methods: either a free vent
to the atmosphere coupled with a flamepath or a 3-mm-thick stainless steel plate. Sufficient safety separation is generally not
achieved by flexible diaphragms.

A free area is generally left open to the atmosphere between the
flameproof enclosure and Zone 0. This buffer region is provided to
avoid flashback from the flameproof housing into Zone 0. This has

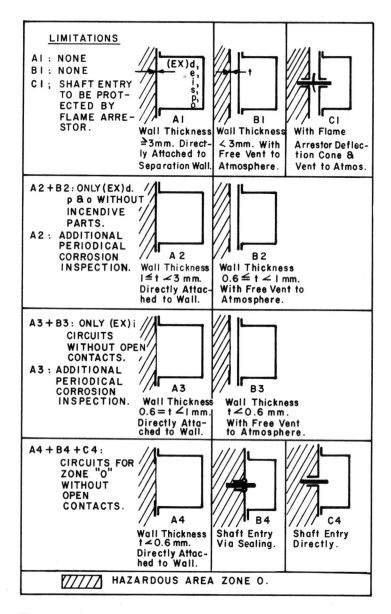

LIMITATIONS

AI : NONE
BI : NONE
CI ; SHAFT ENTRY
     TO BE PROT-
     ECTED BY
     FLAME ARRE-
     STOR.

AI Wall Thickness ≥3mm. Directly Attached to Separation Wall.

BI Wall Thickness < 3mm. With Free Vent to Atmosphere.

CI With Flame Arrestor Deflection Cone & Vent to Atmos.

A2 + B2 : ONLY (EX)d.
     p & o WITHOUT
     INCENDIVE
     PARTS.
A2 : ADDITIONAL
     PERIODICAL
     CORROSION
     INSPECTION.

A2 Wall Thickness I ≤ t < 3 mm. Directly Attached to Wall.

B2 Wall Thickness 0.6 ≤ t < I mm. With Free Vent to Atmosphere.

A3 + B3 : ONLY (EX)i
     CIRCUITS
     WITHOUT OPEN
     CONTACTS.
A3 : ADDITIONAL
     PERIODICAL
     CORROSION
     INSPECTION.

A3 Wall Thickness 0.6 = t < I mm. Directly Attached to Wall.

B3 Wall Thickness t < 0.6 mm. With Free Vent to Atmosphere.

A4 + B4 + C4 :
     CIRCUITS FOR
     ZONE "O"
     WITHOUT
     OPEN
     CONTACTS.

A4 Wall Thickness t < 0.6 mm. Directly Attached to Wall.

B4 Shaft Entry Via Sealing.

C4 Shaft Entry Directly.

HAZARDOUS AREA ZONE O.

Figure 2.10   Combinations of explosion-protective means for level systems with respect to zone classification. Hazardous Classifications are designated in accordance with the International Electrochemical Commission; see Reference 13.

the advantages of reducing the entrance of inflammable gases into the flameproof housing and avoids possible buildup of pressure within the flameproof housing. (Note that the strength of the flameproof enclosure is estimated for the pressure that accumulates as well as the pressure-rise time that takes place during the ignition of the explosive gas mixture). Figure 2.10 illustrates the possible combination of constructions of tank level gauges and their restriction as recommended in Reference 13.

The last member of this hazard group is lightning. Insulated conductors used in Zone 0 should be protected against flashover. Flashover can result when lightning produces sudden potential differences between conductors and other exposed-metal regions.

NOMENCLATURE

b_w Bandwidth

K_c Controller gain

p   Fractional change in controller output

ϵ Fractional change in error

SUGGESTED STUDY PROBLEMS AND QUESTIONS

2.1   Define proportional control and averaging control.

2.2   Explain the purpose of reset control action.

2.3   Name one type of control system used to handle both large and small upsets in a process.

2.4   Explain cascade control.

2.5   A cylindrical tank has a diameter of 300 ft and a height of 85 ft. The vessel is filled at an average rate of 100 gal/min. Compute the vessel's residence time in minutes.

2.6   Define inverse derivative control action.

2.7   A vessel has a holdup of 2500 gal. Compute the rate of level change and new holdup time for a 20% change in the normal throughput to the system. The normal inflow rate is 50 gal/min.

2.8   Compute the time constant for the tank in Problem 2.7, assuming it to be open to the atmosphere and to have an overall height of 80 ft.

3

Visual Techniques for Level Measurement

INTRODUCTION

Visual techniques for detecting the position of process fluid levels are among the oldest and simplest approaches. The simplicity of design for this class of level instrumentation affords a relatively inexpensive and rapid means for level measurement. These systems have no moving parts and so are not subject to mechanical failure. The main disadvantage of visual techniques is that local detection does not lend itself to remote display. Consequently, these devices are limited to level applications involving only minimum exposure to hazardous process conditions. Little basic principle is involved in most of these methods; however, the range of applications to which they are applied is worth noting. The techniques discussed in this chapter are direct gauging devices, manometry methods, and gauge glasses.

DIRECT GAUGING DEVICES

Direct gauging includes a number of very simple devices for obtaining a quick indication of level. The simplest is the common dipstick, which is limited in its length and is used in only nonpressurized vessels. Common applications include liquid level measurement in vessels under atmospheric pressure, obtaining average depths of granular solids in bins, and obtaining local measurements on sludge deposits on tank floors. The types of vessels containing liquids at atmospheric conditions are referred to as burled tanks. The dipstick is simply a scale that is inserted into the vessel, withdrawn, and read. The wetted surface of the scale is an indication of the liquid's height.

The hook gauge, a variation of the dipstick, measures the distance from the liquid's surface to the top of the vessel. The hook, which is immersed in the liquid, is raised to the point where it breaks the liquid surface. It is generally employed in open tanks.

The last device in this group is the tape-and-plumb bob. This device consists of a spooled metering tape with a bob having a specific gravity greater than the fluid being measured. The bob can be dropped through a hatch on the roof of a vessel and as it descends to the tank floor, the tape unwinds. The wetted surface of the metered tape provides an indication of the liquid level. Figure 3.1 illustrates all three devices.

MANOMETRIC PRINCIPLES

A manometer is the simplest technique for measuring differential pressure and thus provides a visual means of detecting the position of a process fluid's surface. The manometer consists of a glass tube bent in the shape of the letter "U" and partially filled with a liquid. When both ends of the U-tube are open to the atmosphere, as shown in Figure 3.2(A), the pressure on each side is equivalent and the column of liquid on one side exactly balances the column of liquid on the other side (that is, the surfaces of the two liquids are at the same level).

The manometer uses the hydrostatic (standing-liquid) balance principle wherein pressure is measured by the height of a liquid it will support. As an example, the weight of a column of mercury at 0°C that is 1-in. high and 1 in. in cross-sectional area is 0.4892 lb. Therefore, a column of mercury 1 in. high imposes a force of 0.4892 lb/in^2.

If one leg of the U-tube is connected by means of tubing or pipe to the bottom section of a vessel or tank, the pressure exerted by the process liquid will cause the manometer fluid to move up the leg of the tube exerting the least pressure. This situation is illustrated in Figure 3.2(B). The response of a manometer to sudden changes in pressure can be described as a second-order system. The basic equation describing this response can be derived from a force balance. For simplification, the frictional pressure drop can be

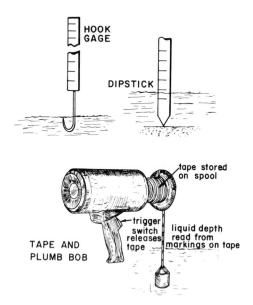

Figure 3.1   Illustrates common direct gauging devices.

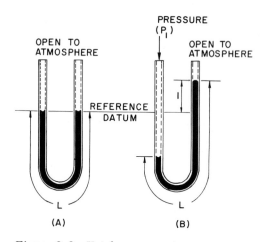

Figure 3.2   U-tube manometer.

assumed to be proportional to velocity and all the manometer fluid
can be assumed to accelerate at a uniform rate. The following
equation describes the manometer's behavior illustrated in Figure
3.2(B):

$$\frac{AL\rho}{g_c} \frac{d^2h}{dt^2} = A\left(P_1 - 2h\rho\frac{g}{g_c}\right) - fA\frac{dh}{dt} \tag{3.1}$$

where   A = cross-sectional area of the U-tube (i.e., flow area of
manometer fluid),

ρ = density of the manometer fluid (it is assumed that the
density of the gas above the fluid can be neglected),

g = gravitational acceleration constant,

t = time,

P_1 = applied pressure,

f = frictional resistance to flow,

L = length of manometer tube filled with fluid, and

h = head of fluid (i.e., height of manometer fluid above
reference mark).

The flow in a capillary tube or manometer is laminar, therefore, the pressure drop across a unit length of the U-tube can be
obtained from the Hagen-Poiseuille equation [17]:

$$\frac{\Delta P}{L} = \frac{32u\mu}{D^2 g_c} \tag{3.2}$$

from whence the frictional resistance to flow is

$$f = \frac{32L\mu}{D^2 g_c} \tag{3.3}$$

where μ = viscosity of manometer fluid,

u = flow velocity, and

D = diameter of the tube.

By substituting Equation 3.3 into 3.1 and rearranging, we obtain

$$\frac{L}{2g}\frac{d^2h}{dt^2} + \frac{16L\mu}{\rho g D^2}\frac{dh}{dt} + h = \frac{P_1 g_c}{2\rho g} = h_i \qquad (3.4)$$

In physics, the classical approach to describing the dynamic behavior of a second-order system (for example, a mass suspended from a spring) is to define the equation in terms of a characteristic damping coefficient. Applying the same principle to Equation 3.4 and defining ω_N as the natural frequency of the system in radians per second and ξ as the damping coefficient

$$\frac{1}{N^2}\frac{d^2h}{dt^2} + \frac{2\xi}{\omega_N}\frac{dh}{dt} + h = h_i \qquad (3.5a)$$

where $\omega_N = \sqrt{\frac{2g}{L}}$, rad/s and $\xi = \frac{8L\mu}{D^2\rho g}\sqrt{\frac{2g}{L}}$ $\qquad (3.5b)$

Solutions to Equations 3.5a and 3.5b can be obtained for step changes in input pressure, using a complete table of Laplace transforms. The value of the damping coefficient will determine the exact form of the solution's response. The damping coefficient can give rise to three situations. For $\xi = 1.0$, the system is said to be critically damped. That is, the manometer fluid (or system's response) arrives at its equilibrium position without overshooting.

When the damping coefficient is zero, the system's response is said to be an undamped sine wave. The frequency of such a system is ω_N and its amplitude is $2h_i$.

For a damping coefficient less than 1.0, the system's output overshoots the equilibrium or final value and oscillates before coming to rest. This system is referred to as being underdamped.

Finally, if the damping coefficient exceeds a value of 1.0, the system is overdamped. In this case, the fluid comes to equilibrium slowly. The solutions to Equation 3.5 for each of these cases are summarized below:

Critically damped system, $\xi = 1.0$

$$\frac{h}{h_i} = 1 - (1 + \omega_N t)\exp(-\omega_N t) \qquad (3.6)$$

Underdamped system, $\xi < 1.0$

$$\frac{h}{h_i} = 1 + \frac{\exp(-\xi\omega_N t)}{\sqrt{1 - \xi^2}} \sin(\omega_N \sqrt{1 - \xi^2}\, t - \phi) \qquad (3.7)$$

where $\phi = \tan^{-1}\left[\dfrac{\sqrt{1 - \xi^2}}{-\xi}\right]$

Overdamped system, $\xi > 1.0$

$$\frac{h}{h_i} = 1 + \frac{1}{\theta_b - \theta_a}\left[\theta_a \exp(-t/\theta_a) - \theta_b \exp(-t/\theta_b)\right] \qquad (3.8)$$

where θ_a and θ_b are time constants obtained from two roots to the solution.

Water- or mercury-filled U-tube manometers are typically underdamped systems. Table 3.1 gives theoretical values of the natural frequencies and damping coefficients for these manometers. Biery [18] has noted that actual values for the damping coefficient are higher than predicted by Equation 3.5. It should be noted that the natural frequency of a manometer typically ranges between 0.5 and 1 cycle/sec. Adjustable damping can be provided by including a valve between the two legs of the manometer. This approach is typically employed when measuring the pressure drop across an orifice with a manometer for flow measurement purposes.

Liquid manometers are widely used in both industrial and laboratory applications. Manometers are unique instruments in that they are both basic pressure measurement devices and standards of calibration of other instruments.

The primary advantages of manometers are that they contain no mechanical moving parts, require no calibration, and are relatively inexpensive compared to other devices. Traditionally the manometer was considered strictly a laboratory instrument. Today they are used to measure pressures ranging from as high as 600 in. of mercury to those as low as space vacuums [19].

Manometers are available in a variety of configurations. It should be noted that only the height of fluid from the surface of one tube to the surface in the other is the actual height of the fluid opposing and balancing the pressure. This statement holds true regardless of the geometry or size of the tubes. Even if the manometer tubes

Table 3.1   Theoretical Damping Coefficients for Water-Filled and Mercury-Filled Manometers

Indicating Fluid	Density (g/cm^3)	Viscosity (cP)	Diameter (cm)	Length (cm)	ω_N (rad/sec)*	ξ *
Mercury	13.5	1.6	0.2	250	2.80	0.169
Mercury	13.5	1.6	0.3	250	2.80	0.075
Mercury	13.5	1.6	0.5	250	2.80	0.027
Water	1.0	1.0	0.2	250	2.80	1.429
Water	1.0	1.0	0.5	250	2.80	0.229

*Values computed from Equations (3.5-a, -b).

are unsymmetrical, the worst that will occur is that more or less
fluid will be moved from one leg to another; however, the fluid
height required to achieve equilibrium will only depend on the density
of the manometer fluid and vertical height. Table 3.2 gives equivalent values of various common indicating fluids. This table demon-
strates the versatility of the manometer. As an example, when
water is used as an indicating fluid, a 10-in. fluid height measures
0.360 lb/in^2. The same instrument utilizing mercury measures
4.892 lb/in^2. This represents a ratio of 13.57:1. This principle is
common to all manometers.

There are three types of pressure measurements that can be
made with manometers:

Positive pressures, or gauge pressures (greater than
atmospheric)

Negative pressures, or vacuums (less than atmospheric)

Differential pressure (the difference between two pressures)

Connecting one leg of the U-tube to a source of gauge pressure,
as illustrated in Figure 3.2(B), causes the fluid in the connected leg
to depress while the fluid rises in the vented leg. However, if the
connection is made to a vacuum, the effect would be to reverse the
fluid movement, causing it to rise in the connected leg and recede in
the open leg.

When measuring differential pressures, one leg is connected to
each of the two pressures. The higher pressure depresses the fluid
in one leg while the lower pressure allows the fluid to rise in the
other. The true differential is measured by the difference in height
of the fluid in the two legs. The manometer also indicates which of
the two pressures is higher (i.e., the higher pressure depresses the
fluid column).

TYPES OF MANOMETERS

Thus far, manometric measurement principles have been discussed
in reference to U-type manometers. Manometers are, however,
available in a variety of forms to provide greater convenience and to
meet different service requirements. Figure 3.3, for example,
illustrates a well-type manometer. As shown in the figure, if one
leg of the manometer is increased in area compared to the other, the

Table 3.2   Conversion Values for Common Manometer
             Fluids at 22°C

---

1 in. water   = 0.0360 lb/in^2
                   0.5760 oz/in^2
                   0.0737 in. mercury
                   1.2131 in. Red Oil

1 ft water   = 0.4320 lb/in^2
                   6.9120 oz/in^2
                   0.8844 in. mercury
                   62.208 lb/ft^2
                   14.5572 in. Red Oil

1 in. mercury   = 0.4892 lb/in^2
                   7.8272 oz/in^2
                   13.5712 in. water
                   1.1309 ft. water
                   16.4636 in. Red Oil

1 oz/in^2 = 0.1228 in. mercury
                   1.7336 in. Water
                   2.1034 in. Red Oil

1 lb/in^2 = 2.0441 in. mercury
                   27.7417 in. water
                   2.3118 ft water
                   33.6542 in. Red Oil

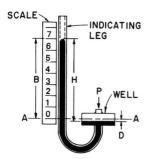

Figure 3.3   Well-type manometer.

volume of fluid displaced represents little change in vertical height
of the large-area leg compared to the change of height in the smaller-
area leg. It thus becomes necessary only to read the scale adjacent
to the single tube rather than two, as in the U-type.

The well design lends itself to use of direct reading scales that
are graduated in appropriate units for the process variable involved
(in this case, level). This manometer configuration does have certain operation requirements. The higher pressure source to be
measured must always be connected to the well leg, while the lower
pressure source must be connected to the top of the tube. This
means that a differential pressure must always have the higher pressure source connected at the well connection. In any measurement,
therefore, the source of pressure must be connected in such a manner that causes the fluid to rise in the indicating tube.

The true pressure follows the same principles previously discussed and the measurement obtained reflects the difference between
the fluid surfaces. Obviously there must be some decrease in the
well level, however, this can be compensated for by spacing the
scale graduations by the proper amount needed to reflect and correct
for the well level drop.

There are a variety of applications requiring accurate measurement of low pressures, such as drafts and low differentials in air and
gas installations. A device commonly used for these cases is the
inclined-tube well manometer, as illustrated in Figure 3.4. This
design has the advantage of providing an expanded scale. As an example, this arrangement permits 12 cm of scale length to represent

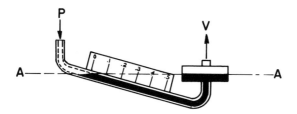

Figure 3.4   Inclined-tube well-type manometer for low pressure
measurements.

1 cm of vertical length.   Using scale divisions of 0.01 in. of liquid
height, an equivalent pressure of 0.000360 lb/in^2 per division can
be detected with water as the manometer fluid.

   To measure relatively high pressures, a longer indicating fluid
tube is required.   Rather than using excessively long manometer
tubes for high-pressure readings, a dual-tube manometer arrangement, as shown in Figure 3.5, can be utilized.   This arrangement
provides for reading the full range of the instrument in only one-half
of the total vertical viewing distance.   The system illustrated in Figure 3.5 consists of two separate manometers mounted on a single
housing.   Manometer (A) is of a conventional well design having a
zero scale at the bottom and graduated upward.   Manometer (B) has
a well and scale zero raised to the 100-in. level with the scale graduated downward.   Connecting a positive pressure source at connections (A) and (B) in Figure 3.5 causes the indicating fluid level to
rise in tube (A) and fall in (B).   It should be noted however that these
are essentially two separate manometers measuring the same pressure with one indicating column rising and one falling.   The fluid
columns pass each other at the 50-in. mark, and as such, it becomes convenient to read the left scale on pressures from 0 to 50 in.
progressing upscale and the scale on manometer (B) from 50 to 100
in. downscale.   Only the lower 50 in. is thus required to read the
entire 100-in. range.   Gauge pressure connections must be made at
connections A and B, while vacuum connections must be made at V
and V'.

   One additional system frequently used is a sealed tube or absolute manometer.   The term absolute pressure originates from the
fact that in a perfect vacuum, the complete absence of any gas is referred to as "absolute" zero.   For an absolute pressure manometer,

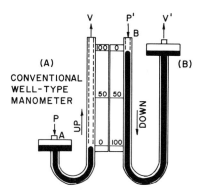

Figure 3.5   Dual-tube manometer system for measuring high pressures; (A) conventional well design; (B) well and scale zero raised to 100-in. level.

the pressure being measured is compared to the vacuum or absolute pressure in a sealed tube above the indicating fluid column. The most common type of sealed tube manometer is the conventional mercury barometer which is used to measure atmospheric pressure. The barometer is a mercury-filled tube over 76 cm in height, immersed in a reservoir of mercury which is exposed to the atmosphere. The mercury column is supported by the atmospheric or barometric pressure. There are a variety of processes, tests, and calibrations that are based on pressures near or below atmospheric conditions. These are most conveniently measured with a sealed tube manometer (referred to as an absolute pressure manometer).

LIQUID LEVEL MEASUREMENT VIA MANOMETRIC TECHNIQUES

The term hydrostatic refers to standing water or liquid. The pressure that is exerted by a liquid per unit area is equivalent to the height of the liquid column of known cross-sectional area multiplied by the weight of the liquid per unit of volume. That is, pressure is merely an expression of force or weight per unit area. Figure 3.6 illustrates the concept of hydrostatic force. Knowing the weight per unit volume of a liquid, the height or depth of that liquid can be expressed as a pressure. Consequently, any pressure measuring

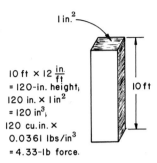

Figure 3.6   Illustrates the principle of hydrostatic force.   10 ft x 12 in./ft = 120-in. height; 120 in. x 1 in^2 = 12 in^3; 120 in^3 x 0.0361 lbs/in^3 = 4.33-lb force.

device having the proper pressure range and sensitivity can be applied as a liquid-level measuring instrument.

Manometers are good candidates for use as tank level gauges. They are attractive because they portray the tank liquid level as a liquid height in the indicating tube. There are a number of ways in which manometers can be implemented for level indication, however, and many of these approaches are unique to specific applications.

One of the simplest approaches is to use a single-tube indicating column or straight U-tube-type manometer directly connected to the tank. This would allow the vessel contents to actually flow into the manometer tube. Such a technique is acceptable for small tanks and for those requiring hard rubber or other special noncorrosive construction. Figure 3.7 shows one design commonly used. Practical limitations may restrict this approach, for example, for larger tanks the instrument must be as long as the tank is high. Additional limitations are that the measuring device must be mounted either on, or close to, the vessel and should the gauge break, the tank contents would leak from the vessel.

Another direct-connection approach is illustrated in Figure 3.8. This method uses a well-type manometer, connected directly to the tank. The system uses an indicating fluid that is heavier than the vessel-content fluid. This has the advantage in that a manometer range shorter than the total tank height can be used.

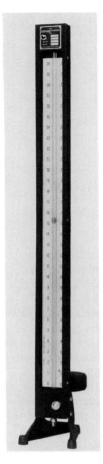

Figure 3.7   U-type manometer mounted in a sturdy steel case with
stainless-steel wetted parts.   The unit is available with duplex,
metric, and special scales.   This unit has a maximum operating
pressure of 250 lbs/in^2.   (Courtesy of Meriam Instrument, Division
of The Scott & Petzer Co. , Cleveland, Ohio).

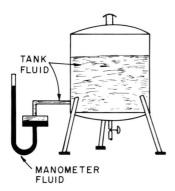

Figure 3.8   Well-type manometer used for measuring level height.

A purge technique has also proven useful in many applications. This approach is illustrated in Figure 3.9 and involves purging air or a suitable gas into the vessel contents through a bubble pipe or tank bell. The gas pressure that accumulates in the bubble pipe is equivalent to the hydrostatic pressure of the liquid. By applying this pressure to a manometer, a remote liquid-level gauge system is created.

Another frequent approach is to employ a conventional pneumatic transmitter to sense the tank level. The transmitter can consist of a flush diaphragm or bubble purge system. The transmitter pneumatic output pressure (gauge) would be sent to a conventional manometer. The span of the transmitter could be adjusted to provide an output range to cover the full tank depth. The approach utilizes manometer tubes having the same length for several vessels of different maximum depths.

Level detection for liquified gases employs still another manometric arrangement. Gases such as carbon dioxide and nitrogen are often stored in the liquid state at low temperatures. These materials turn to a gaseous state at room temperature and manometers can be used to take advantage of this. Figure 3.10 illustrates one arrangement for liquified gas storage tank gauging. As shown, the gas itself serves as the purge.

Multiple manometer tube units are common when gauging a number of vessels. Generally, it is desirable to utilize minimum space for indicating tank levels for several installations. Multiple

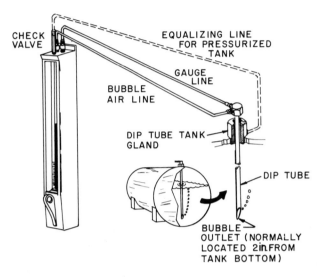

CHECK
VALVE

EQUALIZING LINE
FOR PRESSURIZED
TANK

GAUGE
LINE

BUBBLE
AIR LINE

DIP TUBE TANK
GLAND

DIP TUBE

BUBBLE
OUTLET (NORMALLY
LOCATED 2in.FROM
TANK BOTTOM)

Figure 3.9   Gas bubbler or gas purge system for tank gauging.

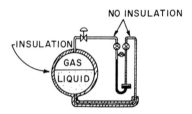

NO INSULATION

INSULATION

GAS

LIQUID

Figure 3.10   Illustrates one arrangement for liquified gas
storage-tank gauging.

tube manometers are available usually in a common support structure. This allows a quick comparison of the levels in all tanks
involved.

Other situations frequently encountered include gauging of
elevated storage tanks and corrosive liquid storage. For elevated
water storage tanks, direct-connecting level gauges can be used.
Such a system is illustrated in Figure 3.11. Note that the height of
interest in Figure 3.11 is H_T ; that is, the gauge should only reflect

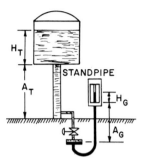

Figure 3.11   Example of an elevated storage-tank gauge installation.

H_T in the tank and not include the standpipe height A_T. To cancel
out the effect of the water in the standpipe, the manometer well is
positioned at some predetermined point, A_G, below the measuring
scale. This canceling effect is referred to as suppression. For the
illustration given in Figure 3.11, the scale indicates H_G equivalent
to H_T.
    When handling corrosive fluids special consideration must be
given to materials of construction and installation of tank gauges. A
typical corrosive fluid delivery and storage system is illustrated in
Figure 3.12. Safety considerations restrict the use of gravity drain
connections on transport vehicles. Normally, a dip tube is installed
in the tank truck or tank car. Air pressure applied to the transport
tank forces the liquid into the receiving storage tank. Storage tanks
are subjected to pressure surges during the fill operation, even
though the vessel is vented. Resulting pressure surges can force the
stored corrosive fluid up into the purge line and to the level gauge,
resulting in corrosion. To avoid this situation, a pressure balancing
line can be used between the storage vessel and the top connection
(i.e., the low pressure connection) on the level gauge. A balancing
line transmits pressure surges to the other side of the level gauge,
thus preventing the corrosive fluid from flowing into the purge system of level gauge.
    There are a variety of indicating fluids available for the different
applications discussed above. In establishing the range of the instrument needed for a specific installation, the following formula can be
used:

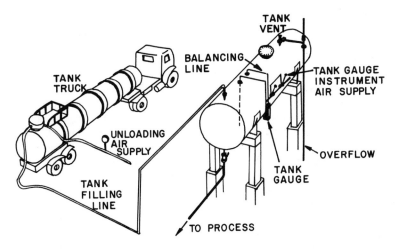

Figure 3.12   Level gauge system for corrosive-fluid storage system.

$$H_G = 12\frac{H_T S_T}{S_i} \qquad\qquad (3.9)$$

where H_G = maximum range of instrument (in.),

H_T = tank depth (ft),

S_T = specific gravity of tank contents, and

S_i = specific gravity of indicating fluid.

Note that the lighter the indicating fluid, the longer the instrument scale. In addition, finer minor scale graduations can be used.

GUIDELINES FOR MANOMETRIC OPERATION AND MAINTENANCE

The manometer should be mounted at a convenient height for vertical reading on the wall, panel, or table, depending on the mounting style. The user should always make sure that the instrument is level and check the fluid level on the side and front of the instrument case.

Most commercial units have an adjustable level that can be used as a
measure of level mounting. Also, most models are equipped with a
scale adjustment. The scale should be positioned at the center of the
adjustment span.

The indicating fluid procedure differs with the design. For U-
tube manometers, the top manometer head or fill plug should be removed. The instrument should be properly vented on the low-pressure side. The selected manometer fluid should be slowly poured
into the glass tube until the indicating fluid level is at approximately
the zero graduation on the scale. Air bubbles should be eliminated
from the tube prior to replacing the head. The scale can then be
adjusted to the correct zero position in relation to the indicating fluid
meniscus. For well-type manometers, the same procedure is followed except that the indicating fluid is poured into the well. For
inclined tube manometers, the proper manometer fluid must be
selected, based on the manufacturer's recommendations.

To obtain consistent measurements, the fluid meniscus should
always be observed in the same manner. For vertical tube devices,
the fluid level at the center of the tube should be noted for each
measurement, whether the top surface is concave or convex in shape.
For inclined tube instruments using mercury, the highest indicated
liquid level, as measured by a line parallel to the graduations of the
scale, should be read. If other indicating fluids are used, the lowest
visible level should be read, as measured by the same procedure.
For U-tube manometers, the levels in both legs must be read and
these readings added together to obtain the actual measurement. An
easy way to read a manometer scale properly is to imagine a plane
tangent to the fluid meniscus and at a right angle to the tube bore
intercepting the scale. Figure 3.13 illustrates the proper manner in
which to read manometer measurements.

Maintenance for manometric systems is relatively straightforward and usually involves occasional cleaning of the tubes. Some
manometer indicating fluids become oxidized after a certain period
of use. Other fluids may react with the various gases and fluids they
contact, producing deposits within the tube; however, such depositing
generally takes a long time to occur. It is simplest to clean the unit
from the upper end of the manometer. For installations consisting
of return wells, liquid checks, etc., at the upper manometer connection, it is generally best to remove this equipment so that the
instrument can be cleaned through the drain plug connection in the
lower channel and block. Whenever possible, the instrument should
be removed and dismantled for cleaning with an appropriate solvent.

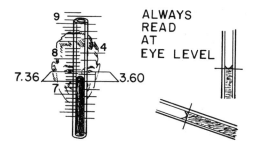

Figure 3.13   Proper manner to read manometer fluid level.

GAUGE GLASSES

Direct-reading level-gauge glasses have been recognized for many
years as a means of accurately determining the level in pressure
vessels, boilers, evaporators, stills, tanks, and the like. Glass
gauges can be used for measuring many kinds of process liquids,
ranging from ordinary water to highly corrosive chemicals.

   Tubular level gauges are the most commonly used among this
class of visual level indicators. Basically, this type of gauge consists of a set of valves and a glass insert. Some designs include ball
checks located in the valves to meet ASME (American Society of
Mechanical Engineering) Boiler Code Safety Standards. The tubular
glasses are usually provided with suitable protector shields which
can be constructed of plastic, wire, glass, or metal. Tubular design glasses generally range in length up to 183 cm.

   Flat glass gauges are also widely used and are generally supplied in multiple sections. These systems are used for service up
to 10,000 lb/in^2 [20]. Figure 3.14 illustrates both the tubular and
flat glass designs.

   Reflex glass, another common glass gauge device, is based upon
the optical law of total reflection of light when passing from a medium of greater reflective power into a medium of less reflective
power. When groove facets are cut at the proper angles in the inner
surface of the glass, it is possible to eliminate all light from the
vacant space back portion of the glass, while at the same time, light
is permitted to pass through the portion covered with the process
fluid. The result is that a sharp, clear line marks the height of the
liquid above which the air or empty space has a bright mirror-like

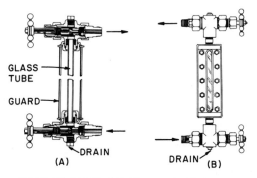

GLASS TUBE

GUARD

DRAIN

(A)

DRAIN (B)

TUBULAR GLASS GAUGE   FLAT GLASS GAUGE

Figure 3.14   Two types of direct-reading glass level gauges.

appearance, while the liquid appears to have the same color as the background in the chamber. The background color is generally black, to provide maximum contrast.

NOMENCLATURE

A   area, ft^2

D   diameter, ft

f   friction factor

g   gravitational acceleration constant, 32.2 ft/s^2 or 980 cm/s^2

g_c conversion factor, 32.174 $(lb_m)(ft)/(lb_f)(s^2)$ or 4.17×10^8 $(lb_m)(ft)/(lb_f) \cdot (h^2)$

H_G manometer range, in.

H_T tank depth, ft

h   head of fluid, ft

h_i initial head of fluid, ft

L   length, ft

P_1 pressure, lb/in^2

ΔP pressure differential, $\mathrm{lb/in}^2$

s_i specific gravity of indicating fluid

S_T specific gravity of tank fluid

t   time, s

u   velocity, f/s

Greek Letters

$\theta_{a,b}$ time constants, s

μ viscosity, $\mathrm{lb_m/ft(s)}$, $\mathrm{lb_m/ft(h)}$, or centipoise (cP)

ξ damping coefficient [see Equation (3.5)]

ρ density, $\mathrm{lb_m/ft}^3$ or $\mathrm{g/cm}^3$

ω_N natural frequency, rad/s

SUGGESTED STUDY PROBLEMS AND QUESTIONS

3.1   Compute several theoretical damping coefficients and
      natural frequencies for different fluid column lengths of
      a manometer using Red Oil as an indicating fluid and
      having a tube diameter of 0.25 cm.

3.2   A U-tube manometer consists of 0.6-cm-diameter tubing
      and has a mercury column 92 cm in length. Each leg of
      the manometer is connected by 300 cm of the same size
      tubing to a pressure tap in a water line. How does the
      presence of water above the mercury affect the damping
      coefficient and critical frequency? Assume laminar flow
      for both the water and mercury.

3.3   Specify the measuring range, in inches, for a U-tube
      manometer to be used on a water tank 60 ft in height.
      Mercury is to be used as the indicating fluid.

3.4   Repeat Problem 3.3, using water as the indicating fluid.
      Which of the two fluids provides the shorter range?

3.5   What three types of pressure measurements are made?

4

Float-Actuated Devices

INTRODUCTION

Float-actuated level measuring devices operate by float movement; that is, a float moves up or down with changes in fluid movement. The float movement can be translated by various means into control action, normally by the positioning of a control valve on the inflow or outflow line to the vessel. Traditionally, float-actuated devices have been limited to the detection of liquid-gas interfaces, and with specially weighted floats, to liquid-liquid interfaces. A variety of commercial designs employed in liquid services are discussed in this chapter. Common to all these devices is the advantage that the level-sensing mechanism is unaffected by the process pressure. These systems are generally employed in applications where glass cannot be tolerated or where a sealed chamber is required.

SIMPLE FLOAT AND LEVER VALVE DESIGNS

One of the simplest types of level controllers is the ball-float-operated lever valve. These systems are employed in a variety of applications in on-off service, ranging from household plumbing to the operating of valves or alarms in such applications as high levels in compressor, fuel-gas knockout, and distillate drums. In general, these devices are employed in situations requiring dependable action at infrequent intervals. Figure 4.1 illustrates two common designs used in liquid service.

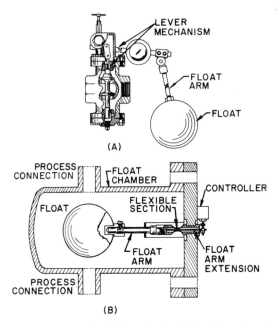

Figure 4.1 (A) Old-style ball-float level control valve for on-off control in open tanks. (B) One type of ball-float level device employed for emergency and alarm services.

The unit illustrated in Figure 4.1(A) consists of a spherical float attached to a lever mechanism which operates a valve member via mechanical levers of either the first- or second-class leverage type. This type of level controller is employed for maintaining the level in a tank by allowing liquid to be fed to the tank at the same rate in which it is being fed to the vessel. Specific applications include building water storage tanks and cooling tower bottoms, where the valve allows makeup water to compensate for evaporative conditions. This particular design is strictly limited to atmospheric systems (i.e., open tank work) and not for pressure vessels. Valves of this design are normally the double-ported, balanced type because the pressure of the supply liquid as furnished to the valve is balanced and, therefore, unbalanced pressure forces are maintained at a minimum. If this were not the case, the float would have insufficient power to operate the valve. It should be noted that double-ported valves generally do not provide tight shutoff and consequently the type of float valve illustrated in Figure 4.1(A) is subject to leakage problems.

The design illustrated in Figure 4.1(B) is also widely used for on-off control. The flattened portion of the ball arm permits sufficient motion to be transferred out of the pressure zone to operate either a pneumatic or electrical switch.

Float-actuated device application is limited because floats are not capable of developing sufficient torque to overcome kinetic unbalance forces. Unbalance forces such as velocity forces result from differential pressure or pressure drop across the valve member [21].

Standard-sized floats are generally of 8-in.-, 10-in.-, and 12-in.-diameters. Large-size floats (12-in.-diameter) are normally used for larger valves (i.e., 4- or 6-in.-diameter). Larger floats are generally not considered practical because of space limitations and cost.

When specifying a float device, the power potential of the float should be checked to ensure that the unit has not been oversized. The valve operation is entirely dependent upon the available power supplied by the ball float. The maximum power that a float can achieve for movement in both directions is realized when the float is submerged 50%. The power potentialities for a ball float of specified diameter can be analyzed quite readily. As an example, consider an 8-in.-diameter ball float for use in a water tank. The cross-sectional area of the float at 50% submergence is

$$1/4\pi\,(8\text{ in})^2 = 50.3\text{ in}^2$$

If under normal operation, the level of the water rises or falls 2 in. above or below the center line of the float, then the volume of the segment of the float corresponding to the level change is

$$2\text{ in} \times 50.3\text{ in}^2 = 100.6\text{ in}^3$$

The density of water is 0.036 lb/in^3, hence, the weight of fluid responsible for the float movement is

$$0.036 \times 100.6 = 3.62\text{ lb}$$

The power available for the operation of the valve is equivalent to the fluid's displacement. Therefore, a 2-in. change in level about the float results in an available power of 3.62 lbf. This force can be transmitted to the inner-valve mechanism through a series of levers. For the valve to travel over its complete cycle, lever movement must be established within a specified range that is proportional to the amount of level change. In the example, if the amount of valve travel required is 2 in., and the level is set to change 10 in., then the operating ratio of the lever system is 5:1. Consequently, for a force of 3.62 lbf from the float, the force transmitted to the valve stem via the lever mechanism is 3.62 x 5 = 18.1 lbf.

This, of course, is an ideal situation as we have assumed a level change on the float of 2 in. and have ignored friction and velocity forces applied by the liquid. In the absence of these dynamic forces, the float would simply follow the level and no displacement of the float would be needed. In real systems, however, these forces are significant and must be overcome. In practice, then, it is desirable to utilize only a small portion of the float through the center such that the smallest amount of level change produces the greatest amount of power.

The applied force determined for the 8-in. float is probably sufficient to operate small valves (2-in. valves or smaller) with pressure drops across the valve as high as a few hundred pounds. For larger valves, the pressure drop will decrease accordingly, giving rise to velocity forces that can exceed the potential float power. This can result in the valve's loss of control.

As valve size is increased, velocity forces are increased in magnitude approximately equivalent to the capacity relationship of the valve. As such, larger valves require larger-sized floats to overcome these forces. If valve and float sizes are not matched accordingly, the inner-valve position can take control of the float position rather than the float having dominance over the valve. To prevent this situation from occurring, a throttling inner-valve design is sometimes employed. These throttling designs generally consist of an internal pilot-actuated lever or float valve.

For the latter designs, the float actuates both the valve stem and a small pilot valve. The advantage of these systems is that the pilot valve only requires a small force to affect movement. The type of float-operated valve is generally single-seated. Figure 4.2 illustrates this type of design. As shown, the supply pressure enters over a valve disk which assists in keeping the valve closed. A composition seat which consists of a soft disk ensures tight shutoff. The supply pressure equalizes across the operating piston through an

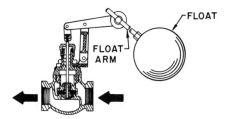

Figure 4.2   Illustrates an internal-pilot-operated ball-float level
control valve unit.

orifice located in the piston. When the pilot valve is in the closed
position, the pressure across the pistons becomes equalized. Also,
the supply pressure over the valve disk keeps the valve closed tight.
When the float signals the valve to open, the valve stem moves upward and the pilot valve opens. Since the area of the piston is larger
than the valve area, an upward force exists that is greater than the
force holding the valve on its seat. This causes the valve-and-
piston system to move upward which in turn opens the valve to a distance that is established by the stem movement. As the piston moves
upward, the stem begins to close the pilot orifice, causing the valve
to come to a balanced situation in which it opens no farther. The
pressure across the piston becomes equalized if the float device then
closes the pilot valve. The valve is then closed by the action of the
supply pressure across the main disk.
      The design illustrated in Figure 4.2 is subject to fouling. Par-
ticulates such as sand or grit can become lodged behind the piston,
rendering the valve inoperative.

CHAIN OR TAPE FLOAT GAUGE DEVICES

These designs consist of a float connected by means of a flexible
chain or tape to a rotating member. This rotating member is in turn
connected to the indicating and control device. Figure 4.3 illustrates
the operational concept. As shown, a counterweight is employed to
maintain the chain or tape in a taut condition while the float rises and
falls. With chain-type gauges, the chain is designed to engage a
sprocket which turns the rotating member. With the tape-type systems, the tape wraps around a drum.

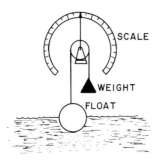

Figure 4.3   Illustrates the operational concept behind chain or
tape-type float gauges.

        These systems can either be installed within a vessel or in a
long section of pipe positioned adjacent and connected to the tank by
suitable connections into the liquid and vapor phases of the vessel.
Figure 4.4(A) illustrates one manufacturer's tape unit in operation.
The float for this system is a heavy-duty, circular steel-foam glass
float [i.e., foam glass which is enclosed in a 316 stainless steel (SS)
jacket]. The float acts upon a counterbalanced, nongraduated, perforated tape which actuates a dial-counter. The gauge is gastight and
is powered by a motor that maintains constant tension over the entire
deflection. The dial-and-counter reading minimizes the possibility
of error inherent in the reading of older designs based on graduated
tapes. The counter assembly for this unit is independent of the main
gauge-head housing. This allows independent adjustment of the
counter assembly without the need for entering the main housing [22].
Systems such as this can be mounted either at the bottom of the vessel as a conventional ground reader, with piping coming in from the
top of the gauge head, or at the top of the tank with the piping entering from the bottom.
        Simpler indicating mechanisms are available for applications
where extreme accuracy is not critical and cost is of prime importance. Figure 4.5 illustrates the same float-actuated level device
operating a gauge board.
        The devices discussed above can be equipped with guided or un-
guided floats, and high- and low-level limit switches to actuate signal lights, horns, relays, or solenoids controlling pumps, motorized
valves, and other operating equipment.

(A)

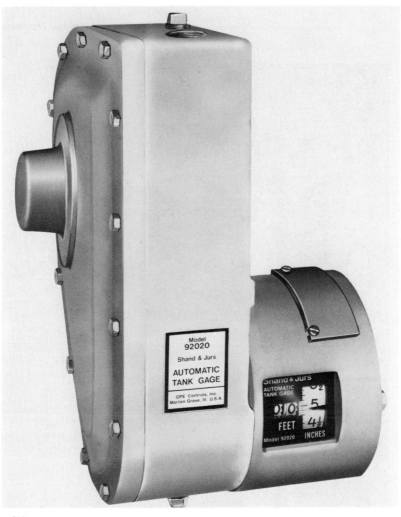

(B)

Figure 4.4 (A) Illustrates a tape-type float gauge in operation.
(Courtesy of VAREC Div., Emerson Electric Co., Gardena, California). (B) Shows a ground-reading tank gauge operated on the
tape-float principle. A perforated stainless steel tape rotates an
aluminum sprocket wheel with stainless-steel pins, which in turn
drives the digital readout counter. A constant-force motor maintains
tension on the float throughout the gauging range. Remote reading
can be added by removing the counter housing cover and installing
a coupling and transmitter in its place. (Courtesy of GPE Controls,
Inc., Morton Grove, Illinois).

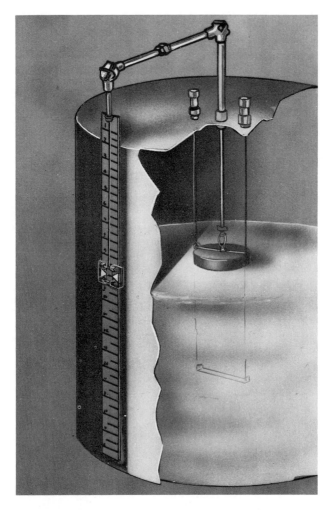

Figure 4.5   Chain-type unit in operation. (Courtesy of VAREC Div.,
Emerson Electric Co., Gardena, California).

MAGNETIC-TYPE FLOAT SYSTEMS

These systems make use of magnetic forces to detect float deflection. Two basic designs are magnetic bond devices and magnetically
operated float switches.

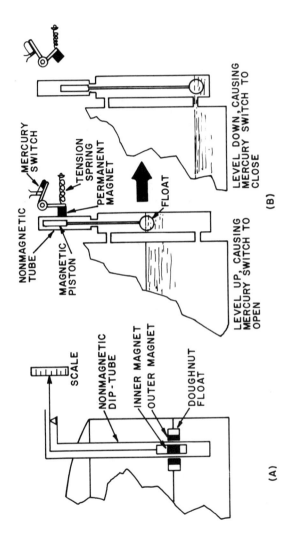

Figure 4.6 Illustrates operation of magnetic-type float methods. (A) Magnetic-bond level controller.
(B) Magnetically operated float switches.

With the former system, level detection is made possible by a magnetic member floating on the surface of the liquid. The magnetic flux field generated is then transmitted from the magnetic float to a suitable detection arrangement. The detector or signal pickup device actuates the controlling device. Figure 4.6(A) illustrates a dough-nut-type float in operation. The float design is not limited to this particular geometry but can be spherical, disk-like, etc., in shape. Note that the dip tube must consist of a nonmagnetic chamber, which is generally constructed of 304 SS pipe.

Figure 4.6(B) illustrates the operation of the latter method—that is, a ball-float mechanism which employs magnetic force to operate a mercury switch. This design operates as follows. A ball float positions a magnetic piston connected to the float arm, which moves up and down within the nonmagnetic enclosing chamber. Outside the enclosing chamber is a permanent magnet connected to a pivoted arm mounted with a mercury switch. When the level goes up, the magnetic piston enters the magnetic field and is drawn against the enclosing tube. This action causes the mercury switch to tilt in one position and opens or closes a circuit, as specified. When the float falls to some predetermined position, the piston moves downward, out of the magnetic field. Thus the magnet is pulled out by the tension spring, causing the mercury switch to tilt in the opposite position.

FLOAT CAGES

In many applications it is neither practical nor desirable to install a float in the tank. In these situations, float cages are applied to the vessel. There are four ratings for float cages, namely, 125-lb and 250-lb iron, and 600-lb or 1500-lb steel.

The simplest type of float cage consists of a dip tube or pipe in which the float is mounted. The float cage can be mounted through the vessel as illustrated in Figure 4.7, or mounted externally. As the level in the tank changes, the same change is transmitted into the float cage since the two levels must be equalized. Changes in level result in changes in the float position, as the float rides on the level in the cage. As illustrated earlier, the float motion is converted into work to operate a control valve.

The internal chamber illustrated in Figure 4.7 is analogous to a manometer with unequally sized legs. From the force balance derived for a manometer [Equation (3.1)], an equation for the natural

Float Cages   73

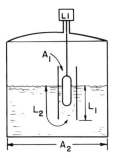

Figure 4.7   Float cage of internal mounting design.

frequency of this system can be obtained:

$$\omega_N = \sqrt{\frac{g(1 + A_1/A_2)}{L_1 + L_2 A_1/A_2}} \qquad (4.1)$$

where A_1, A_2, respectively, are the cross-sectional areas of the
float cage and vessel, and L_1 and L_2 are distances defined in
Figure 4.7.

If the tank is very large in comparison to the float cage (i.e.,
$A_1/A_2 << 1$) then Equation (4.1) can be approximated by

$$\omega_N \simeq \sqrt{\frac{g}{L_1}} \qquad (4.2)$$

As an example, a cage length of 10 ft has a natural frequency of
1.8 rad/s. It is not possible to predict the precise damping coefficient for this system, due to entrance losses and the existence of
turbulent velocity gradients. The damping coefficient would, however, be quite small (see Table 3.1 for examples).

Float cages of the external-mounting-type design consist of two
separate connections through the outside of the vessel. One connection runs from the bottom of the cage into the liquid in the tank. The
other connection runs from the top of the cage into the vessel's vapor
space above the liquid level. Again, as the level in the tank changes,
the same change is transmitted into the float cage, since the two
levels must be equalized. An external float cage and lever valve sys-

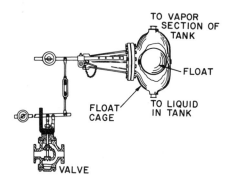

Figure 4.8   External float cage and lever valve.

tem is illustrated in Figure 4.8. Connections screwed to the tank
are adequate for many applications; however, in cases where high
pressures and/or temperatures exist, cage connections should be
flanged. Caution should be exercised in specifying the proper connection size. If the connections are too small, the level in the cage
can lag behind the level in the tank. This can force the system into
a continuous cycling or hunting action since the controller will be out
of phase with the actual level in the tank.

For the system illustrated in Figure 4.8, the float cage unit is
used to operate lever valves in direct-connected assemblies. This
type design requires the valve to be located as close as possible to the
float cage. Float cages of this design can be used to cause level
changes to operate switches to start motors, actuate alarms, or to
signal and indicate level positions remotely.

The natural frequency of external cages depends on the length and
size of connecting pipe and the dimensions of the cage. Harriott [2]
and others [23] have derived expressions for the natural frequency of
external cages based on the manometer analogy. Also, experimental
damping ratios have been reported for water and other low viscosity
liquids [24, 25].

External cages have the advantage that they can be easily modified to damp oscillations by the installation of a valve in the connecting line. If damping is required for an internal cage system, either
a restriction at the bottom of the cage would have to be added or the
unit be replaced by a perforated cage.

REMOTE LEVEL CONTROL DEVICES

Pilot-operated liquid level controllers provide a means by which the primary change of a level is transmitted through the action of the float (or, in general, by other devices discussed in subsequent chapters) to an amplifying relay that requires only small motive power. The amplifying medium transmits an energizing medium to actuate a control valve or some other final control element.

These devices have the advantage of remote installation, i.e., the controller or measuring arrangement can be situated within the vessel at a remote distance from the control valve.

Pilot-operated control mechanisms require only small increments of power output from the float to cause the pilot to transmit an output pressure change. The pilot or relay has an amplifying action which allows for magnification of power development at the primary control valve. This ensures its positive placement at any given position.

Ball-float-type, pilot-operated controllers utilize air or hydraulic pressure transmitted from the pilot valve to a diaphragm-actuated control valve. Pressure variations acting on the large diaphragm and spring arrangement of the valve cause a positive movement of the valve for each pressure condition signalled by the pilot member. Figure 4.9 illustrates an air-pilot-actuated, ball-float level controller arrangement.

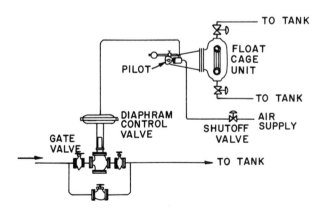

Figure 4.9   Illustrates an air-pilot-actuated float-level controller. The float cage would be mounted directly on the vessel shell.

AUTOMATIC TANK GAUGING AND INVENTORY CONTROL
SYSTEMS

The simple lever-valve mechanism is one of the oldest designs of
float-actuated devices, and although still employed, it has largely
been replaced by more sophisticated and accurate control arrangements. Many of the newer variations still operate on very simple
principles. Figure 4.10 illustrates one level measuring device
equipped with remote readout. The unit performs a level measurement when an operator activates a remote control unit. The sensor
unit mounted on the top of the tank lowers the float into the container,
generating pulses as it travels downward. When the float or probe
impacts the liquid interface, a change in the cable tension stops the
measurement pulses. When this happens, the motor reverses and
returns the float to the top. The pulses for the unit shown are
generated at a rate of 100 per foot by passing the cable over a measuring wheel, and are stored in a digital counter in the control unit.
Although a system like this is designed to signal a flow controller
automatically, it has found its widest application in automatic inventory registering. These systems are employed for work on round
vertical tanks and above- or below-ground horizontal tanks. There
are no dimension limitations for tanks to restrict their use.

Measurement accuracy is a critical consideration in selecting
inventory control systems for large-volume applications such as tank
farms. As an example, consider a gasoline storage tank having a
diameter of 80 ft. Each 0.01 ft of tank height has a volume of
50.27 ft^3. Assuming an average density of 42 lb/ft^3, each 0.01-ft
segment of tank height contains 211 lb or 26.37 gal. If the measurement accuracy of the level indicator is only good to within ±2%, an
error of over 1000 gal would be realized. Consequently, accurate
measurement is highly desirable.

Another important consideration in establishing inventory control
systems concerns the choice of the units to be used for data recording. Many liquid inventories are maintained in units of mass (thousands of pounds or kilograms). Consequently, the detection system
must be equipped with a temperature-compensator arrangement to
make necessary corrections for variations in mass contraction or
expansion. Liquid density changes with temperature variations can
be significant in many applications. There are a variety of data recording systems commercially available. Figure 4.11 illustrates
one such controller that automatically records inventories, indicating
day and time. The inventory can be conducted in any desired unit of

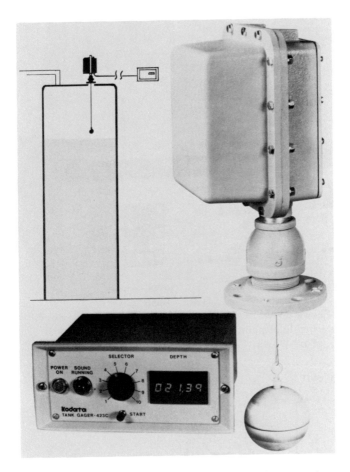

Figure 4.10 Shows an automatic liquid-level measuring system for inventory control. The accuracy of level measurement with this device is repeatable to within 3.05 mm over full depth. (Courtesy of Kodata Inc., Fort Worth, Texas).

measurement (e.g., tons, pounds, gallons, liters, kilograms, etc.). The mass of the inventory is also recorded. These systems have optional temperature recording to provide data for specific gravity adjustment and compensation. Figure 4.12 illustrates a typical arrangement for a tank farm installation.

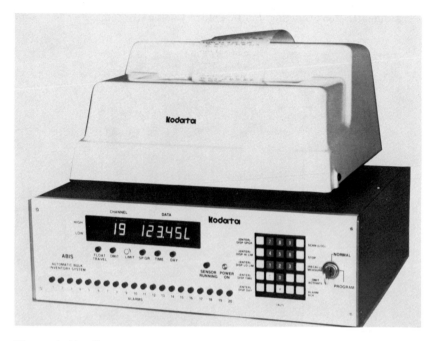

Figure 4.11   Shows an automatic data recorder for tank inventory. The printer is capable of printing 75 lines per minute. (Courtesy of Kodata Inc., Fort Worth, Texas).

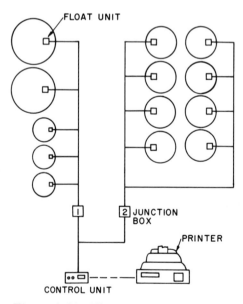

Figure 4.12   Illustrates a typical automatic inventory system for a tank farm installation.

Figure 4.13   Shows one manufacturer's remote storage tank gauging and inventory control system. (Courtesy of GPE Controls Inc., Morton Grove, Illinois).

Figure 4.13 illustrates still another remote storage-tank gauging and inventory control system. This unit is capable of displaying up to 20 tanks simultaneously, as selected by keyboard command. The system continuously displays current level and temperature data and an operator can monitor the display and operation even while the tank is being filled, emptied, or is undergoing temperature change. Systems such as this can be programmed for high- and low-level alarm points for each tank.

SUGGESTED STUDY PROBLEMS AND QUESTIONS

4.1   Derive an expression for the natural frequency of a float cage of the external type. Assume the cage to be a vertical standpipe having a cross-sectional area A, and tank connections each having cross-sections A_2. Assume the float motion to be negligible.

4.2   A cylindrical tank with an overall depth of 15 ft is normally half full. The tank uses a level transmitter with a range of 15 ft and a controller gain such that the vessel would be empty at 300 gal/min and full at 450 gal/min. Determine the cross-sectional area of the tank.

4.3   A 12-in. ball-float level control valve is used to regulate a slop tank containing a 50-50 mixture of oil and water. Under normal operation, the level rises and falls 4 in. above and below the center line of the float. What is the float force transmitted to the lever mechanism? The operating ratio of the lever system is 4:1.

4.4   Repeat Problem 4.3 for an 8-in. float.

4.5   What are the advantages of an internal, pilot-operated, ball-float level control valve arrangement?

5

Probe-Actuated Devices for Solids Level Control

INTRODUCTION

There are a variety of process operations in the mining, metallurgical, pulp and paper, and food industries that require accurate level measurement and control of bulk solids. There are a number of commercial devices successfully employed on blending bins, supply bins, scale hoppers, packaging equipment, etc. This chapter provides descriptive information on specific probe-actuated devices. There are four classes of these devices, namely bob-and-cable tension systems, tilt-switch devices, vibrating or oscillating probes, and rotating shaft-and-paddle systems. As with float-actuated devices, the level indicator or probe must make direct contact with the tank contents.

BOB-AND-CABLE TENSION METHOD

These systems closely resemble float-actuated devices; in fact, their control devices can be used for liquid level service by simply changing the float. Figure 5.1 illustrates the versatility of this type of system. The system shown consists of a plumb-bob-type level indicator that can be used for either liquid or solid services. The unit has a storage-reel design and can be equipped with an electronic high-resolution depth or tank quantity readout (see Figure 5.2). The unit can be mounted atop a silo, tank, or bin and may be connected to a console located either in close proximity or in an area away from the tank. A readout is triggered by a control on the console which

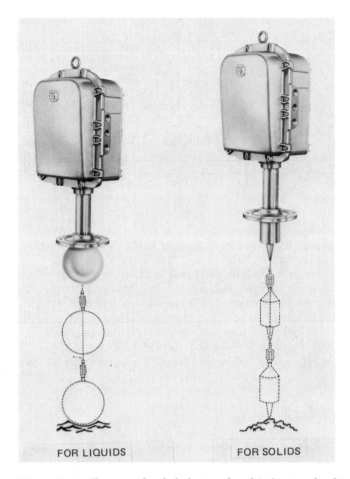

FOR LIQUIDS   FOR SOLIDS

Figure 5.1   Shows a plumb-bob-type level indicator for liquids or
solids. (Courtesy of Monitor Manufacturing, Elburn, Illinois.)

releases a weighted probe supported by a cable. As the probe drops
downward into the tank, a counter in the console provides a continuous reading on the depth. When the probe reaches the tank contents
(e.g., stored liquid, chemical, plastics, cement, coal, paper pulp,
petroleum product, or grain), a slack in the cable is detected by a
sensing device which then reverses the motor to return the probe to

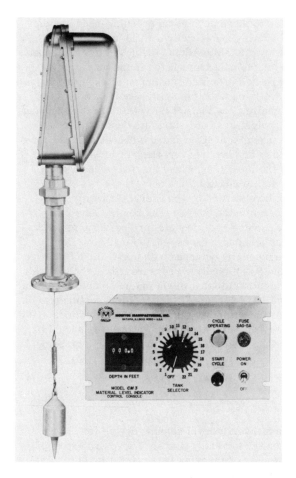

Figure 5.2   These units can be equipped with an electronic counter calibrated to provide a direct readout of the depth of the tank contents.   (Courtesy of Monitor Manufacturing, Elburn, Illinois.)

the top of the tank in its original start position.   The counter in the console maintains the depth reading until another measurement is needed.   Probes can be teflon-coated for special applications involving corrosive fluids.   The unit is not affected by temperature change, humidity, or dusty atmosphere [26].

Manufacterers refer to these devices as sounding-head systems because of the use of a sound-disable switch. The unit is at rest when the sound-disable switch is closed and the shut-off switch open. The "sounding weight" (plumb bob) is socketed in the bottom of a mounting flange and there is a specified tension on a takeup spring. The operating cycle is initiated when the unit is energized by an ac power switch and by signalling the manual trip switch in the sounding head. As the sounding weight or probe starts its descent, the tension is released on the takeup spring, causing a decrease in the tension on the weight-sensing arm spring. As the unit moves downward, the cable drives the measuring wheel and pulse-count cam, sending a signal to the digital counter.

Cable lengths must be specified by the user. Generally, the cable should be sized to match the overall tank height, less 2 ft. Uncoated stainless steel cables are adequate for operating temperatures over 200°F, or for lengths in excess of 100 ft [27].

These systems are normally equipped with a 1/4-in. NPT purge fitting to permit introduction of positive pressure within the unit as a means of excluding foreign matter that might enter the unit during an operating cycle. In addition, this helps to maintain operating temperatures at acceptable limits in high-temperature applications. Any inert gas can be used for purging. Generally, pressures should be at least 1 lb/in^2 higher than the ambient pressure in the tank being monitored.

TILT SWITCHES

Tilt switches have been a favorite level sensing device for bulk solids in the mining industry for many years. These are extremely simple devices and have only one moving part consisting of a steel ball housed within the unit. The unit is normally suspended vertically within the metered vessel by a chain, rope, or rigid hanger to the point of material accumulation. When the tank contents begin to accumulate beyond the specified depth, the unit becomes tilted, causing the internal ball to roll off center. The rolling ball then actuates a mercury microswitch which sends a signal to a recorder, alarm, or control mechanism. Figure 5.3 gives a cutaway view of one type of heavy-duty tilt switch used for large, lumpy materials such as coal, ore, rock, aggregate, or sand. These systems were originally developed for detecting high material level but can be adapted to

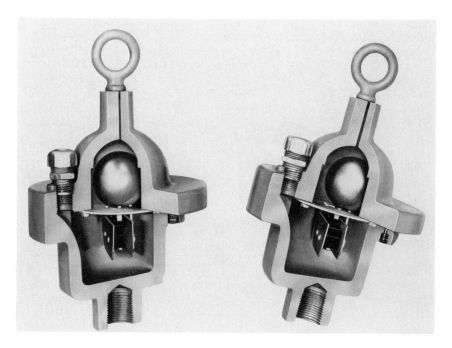

Figure 5.3   Cutaway view of a tilt switch. The steel ball is the only moving part which rolls off center when the unit is tilted, actuating a microswitch for control action. (Courtesy of Monitor Manufacturing, Elburn, Illinois.)

control flow or the operation of other equipment. Normally, a shaft or short length of pipe is added to the base of these units to increase the deflection point.

Figure 5.4 shows a lightweight device for use in bins, silos, and storage tanks, for measuring a variety of solid materials such as grain, plastic pellets, feed pellets, pet foods, fertilizers, and other lightweight materials. This particular type of unit has also been used for control in determining over- or underload on conveyor belts.

Tilt switches can be operated in areas which exclude many conventional controls because of large material sizes and weights or the presence of hazardous atmospheres. In many applications these systems are designed for explosionproof requirements.

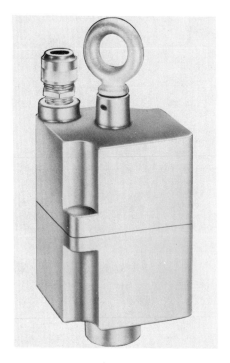

Figure 5.4 Shows a lightweight tilt switch used for low density materials such as fertilizers, feed pellets, etc. (Courtesy of Monitor Manufacturing, Elburn, Illinois.)

OSCILLATING- AND VIBRATING-PROBE LEVEL CONTROLS

Oscillating- and vibrating-probe level controllers are relatively new devices. The unit shown in Figure 5.5 is an oscillating probe device used for controlling material flow in small hoppers, cyclones, pack-aging machinery, and loading equipment common to the plastic and chemicals industry. The unit's shaft and spherical probe are nor-mally mounted through the side of the hopper. There is no live seal between the actuating probe and the drive mechanism in operation. The probe oscillates on its own elastomeric diaphragm which is the pivot point. When incoming material interferes with the travel on the sensing probe, it stops by means of a floating mechanism within the unit; this actuates a switch to energize a desired control signal.

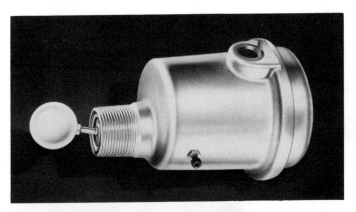

Figure 5.5 Shows an oscillating-probe bin-level controller for solids. The unit has a special threaded cover which unscrews to reveal the inner mechanism. The machined cover and base flanges provide a positive seal against the tank environment. (Courtesy of Monitor Manufacturing, Elburn, Illinois.)

The unit shown in Figure 5.5 was designed to provide a visual signal for a stop-or-go situation or to trigger related operations via an intermediate relay. On bagging setups where the hopper has a gauged unloading rate, should an obstruction of the outgoing material occur, the high-level unit would stop the flow of material into the hopper.

Vibrating-probe devices make use of an integrated frequency response to determine the presence or absence of either liquids or solids in storage vessels. One such unit is shown in Figure 5.6. A

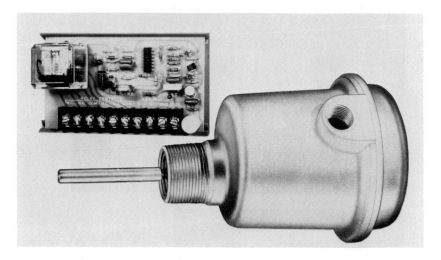

Figure 5.6   Solid-state, integrated frequency bin-level control device. (Courtesy of Monitor Manufacturing, Elburn, Illinois.)

control module sends a pulse to the sensor head. The pulse applies a force to the sensor rod, resulting in its vibrating if no material restricts the sensor rod inside the bin. After pulsing the sensor rod electronically, the control module monitors the vibrations. If the vibrations at both ends of the sensor are the same, no restrictions exist and the material is not present. If the vibrations are unequal, then material is present at the other end of the sensor rod and a relay is triggered. Contacts of the control module can actuate alarms, indicator lights, or various process control equipment.

ROTATING SHAFT-AND-PADDLE DEVICES

These devices are basically mechanical pulse-switch indicators that resemble a conventional mixer assemblies. Figure 5.7 illustrates one design. The unit can be installed through the top, or through the side of a vessel at any intermediate height. The operating principle is simple; as long as the shaft of the level control turns freely, no material is present. When the shaft stops turning, the controller signals an appropriate indicator (alarm, indicating light, etc.) that material is present.

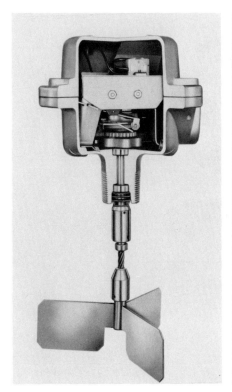 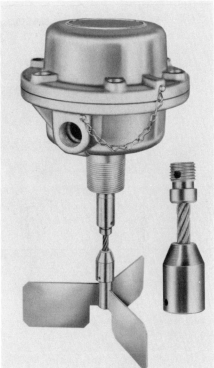

Figure 5.7   Shows a rotating paddle bin-level control. These systems resemble standard mixers. (Courtesy of Monitor Manufacturing, Elburn, Illinois.)

These systems can be adapted to a variety of paddle configurations (See Figure 5.8). Paddle configuration is extremely important
to proper operation (i.e., there is no universal paddle design ideally
suited for bulk materials with a wide range of specific gravities).
The proper selection of a paddle configuration can make a difference
between a successfully operated installation of paddle-type controls
and an unsuccessful one. Major considerations in selecting the proper paddle configuration are material density, solids size, and material flow characteristics [28].

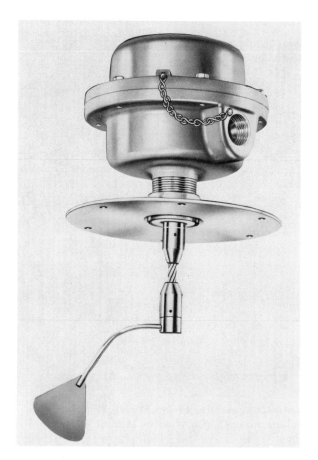

Figure 5.8   Rotating paddle devices are designed to be interchanged with a variety of paddle configurations.   Paddle configuration must be specified in accordance with the materials' properties.   (Courtesy of Monitor Manufacturing, Elburn, Illinois.)

As a rough guide to proper paddle configuration selection for different dry bulk materials, Tables 5.1 and 5.2 and Figure 5.9 can be used.   Table 5.2 gives a partial listing of common dry bulk materials divided into two general groups: minerals and chemicals, and food and agricultural products.   The table gives the material's common name, average density, and a brief description of material

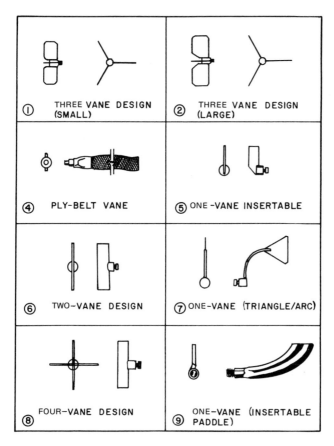

Figure 5.9   Various paddle configurations. The paddle configuration numbers given in Table 5.1 correspond to the drawing numbers of this figure.

properties. Following each material is a recommended paddle code number for high, intermediate, and low level control corresponding to a drawing of the paddle in Figure 5.9. Table 5.2 is the key to the handling properties column in Table 5.1. Material handling properties (Table 5.1) are noted by means of a letter, a number, or combinations of both, the key to which is summarized in Table 5.2.

Table 5.1   Materials Classification and Paddle Configuration Guide (Courtesy of Monitor Manufacturing Co., Elburn, Illinois).

Minerals and Chemicals	Density (lb/ft³)	Properties	Paddle Configuration for Intermediate and	
			High Level	Low Level
Adipic acid	45	B26P	1	5
Alumina, fine, granulated	55	B28	1	5
Aluminum chips	7-15	H36X	2	7
Ammonium perchlorate	52	B26	1	5
Ashes, coal, dry, 1/2 in. and under	35-40	C37	1	7[b]
Ashes, coal, dry, 3 in. and under	35-40	D37	8[a]	5
Ashes, coal, wet, 1/2 in. and under	45-50	C37PZ	8	5
Ashes, coal, wet, 3 in. and under	45-50	D37PZ	8[a]	5
Asphalt, crushed, 1/2 in. and under	45	C26	1	5
Barite	180	D28	6	5
Barium carbonate	72	A37	6	5
Bark, wood, refuse	10-20	H37X	2	7[b]

Bauxite, crushed, 3 in. and under	75–85	D28	6[a]	5
Borax, fine	75	B26	6	5
Calcium, carbide	70–80	B27	6	5
Calcium carbonate (limestone)	85	C26	6	5
Calcium oxide (lime)	27	B26	1	7[b]
Carbon black, pelletized	20–25	B16TZ	2	7[b]
Carborundum, 3 in. and under	100	D28	6[a]	5
Casein, granular	38–43	B27	1	9
Cast iron chips	130–200	C37	6	5
Cellulose acetate	10	H36	2	2
Cement clinker	75–80	D28	6	5
Cement, portland	65–85	A27Y	6	5
Chalk, 100 mesh	70–75	A37YZ	6	5
Charcoal	18–25	D37T	2	7[b]
Chrome ore	125–140	C28	6	5
Cinders, blast furnace	57	D38	1	5
Citric acid	55	B26	1	5

Table 5.1 (Continued)

Minerals and Chemicals	Density (lb/ft³)	Properties	Paddle Configuration for Intermediate and High Level	Low Level
Clays				
attapulgus	55	A27	1	5
bentonite	51	A26Y	1	5
calcined	80	A27	6	5
dicalite	50	A27	1	5
kaolin	64	A27	1	5
sno–brite	50	A27	1	5
whitex	50	A26	1	5
Coal, anthracite	50	C27P	1	5
Coal, anthracite, river or culm, 1/8 in. and under	60	B37P	1	5
Coal, bituminous, mined, 50 mesh and under	50	B36P	1	5
Coal, bituminous, mined, run of mine	50	D26P	1	5
Coal, bituminous, mined, sized	50	D26PT	1	5

Coal, bituminous, mined, slack, 1/2 in. and under	50	C36P	1	5
Coal, bituminous, stripping, not cleaned	50	D37P	1	5
Coal char	24	B37SY	2	7[b]
Copper ore	120-150	D28	6	5
Ebonite, crushed, 1/2 in. and under	65-70	C26	6	5
Feldspar, crushed	75-100	B27	6	5
Feldspar, pulverized	50-60	A27	1	5
Fluorspar	82	C27	6	5
Fuller's earth	35-40	B27YZ	1	7[b]
Gilsonite	37	C27PS	1	9
Glass batch	90-100	D28	6	5
Gravel, screened	90-100	D27	6	5
Gypsum, calcined	55-60	C27	1	5
Gypsum, lumps	82	D27	6	5
Gypsum, powder	60-80	B37	1	5
Ilmenite ore	140	B28	6	5
Iron ore	125-150		6	5

Table 5.1 (Continued)

Minerals and Chemicals	Density (lb/ft³)	Proper- ties	Paddle Configuration for Intermediate and	
			High Level	Low Level
Lignite, air-dried	45–55	D26	1	9
Lime, briquette	60	D27	1	5
Lime, burned, pebble (sized)	53	D26	1	5
Lime, burned, pulverized	27	A26	2	7[b]
Lime, burnt, quick, crushed	50	C27	1	5
Lime, fine, spent, dry, carbide	45	A27	1	5
Lime, ground 1/8 in. and under	60	B26Z	1	5
Lime, hydrated	10–25	B26Y	2	7[b]
Lime, hydrated (acetylene process)	32–40	B26YZ	1	7[b]
Lime, hydrated, 1/7 in. and under	40	B26YZ	1	7[b]
Lime, mason	17	A36	2	7[b]
Limestone, coarse, sized	98	D27	6	5
Limestone, dust	75	A26	8	5
Limestone, mixed sizes	105	D27	6	5

Limestone, pulverized	85	A27WZ	8	5
Litharge	85	B37R	6	5
Magnesium sulphate (epsom salts)	40-50	B26	1	5[b]
Manganese dioxide	80	B27	6	5
Manganese sulphate	70	C28	6	5
Marble, crushed, 1/2 in. and under	90-95	D28	6	5
Mica, ground	13-15	B27	2	7[b]
Mica, pulverized	13-15	A27Y	2	7[b]
Mica flakes	17-22	B17WY	2	7[b]
Muriate of potash	77	B28	6	5
Nitrate of soda, granular	68	C26	6	5
Nylon chips	38	C26	1	5
Phosphate, dicalcium, granulated	60	C27	1	5
Phosphate, Florida, No. 20 mesh	93	B27	6	5
Phosphate, momo-calcium powder	61	A27	1	5
Phosphate, powdered	60	A27	1	5
Phosphate, rock, crushed	75-85	D27	6	5

Table 5.1 (Continued)

Minerals and Chemicals	Density (lb/ft³)	Proper-ties	Paddle Configuration for Intermediate and High Level	Low Level
Phosphate, rock, ground	90	D27	6	5
Phosphate, super, ground	51	B27	1	5
Phosphate, trisodium, powder	50	A27	1	5
Phthalic anhydride, flakes	42	D26	1	7[b]
Polyethylene pellets	34	C26	1	7[b]
Polyethylene, powder	45	A26Y	1	5[b]
Polystyrene, beads	45	A26Y	1	5[b]
Polystyrene, powder	42	A26YZ	1	5[b]
Polyvinyl chloride, pellets	36	C26	1	7[b]
Polysinyl chloride, powder	45	A26Y	1	5[b]
Potassium chloride	75-80	B26Y	1	5
Potassium nitrate	76	C17P	1	5
Potassium sulphate	42-48	B27	1	5[b]
Pyrites, pellets	120-130	C27	6	5

Salt (NaCl) granulated, dry, coarse	45–50	C26PL	1	5[b]
Salt (NaCl) granulated, dry, fine	70–80	B26PL	1	5
Salt cake (Na_2SO_4) dry, coarse	85	D26	6	5
Salt cake (Na_2SO_4) dry, fine	65–85	B26	6	5
Salt, rock, coarse	75	B26	6	5
Salt, rock, ground	80	B26	6	5
Sand, bank, damp	110–130	B38	6	5
Sand, bank, dry	90–110	B28	6	5
Sand, foundry, prepared	90	B38	6	5
Sand, foundry shakeout	90	D28	6	5
Sand, silica, dry	90–100	B18	6	5
Shale, crushed	85–90	C27	6	5
Silica flour	80	A38	1	5
Silica gel	40–45	B28	1	5[b]
Slag, furnace, granular	60–65	C28	1	5
Slag, furnace, lumpy	160–180	D38X	6	5
Slate, crushed, 1/2 in. and under	80–90	C27	6	5

Table 5.1 (Continued)

Minerals and Chemicals	Density (lb/ft³)	Properties	Paddle Configuration for Intermediate and High Level	Low
Slate, ground, 1/8 in. and under	82	B27	6	5
Soap beads or granules	25	B26T	2	7[b]
Soap chips	10-25	C26T	1	7[b]
Soap powder	20-25	B26T	2	7[b]
Soda ash, light	20-35	A27W	2	7[b]
Soda ash, dense	55-65	B27	1	5
Sodium chloride	45-50	C26PL	1	5
Sodium nitrate	70-80	C16	6	5
Sodium phosphate	50-60	A26	1	5
Sodium sulphate	45-50	B26	1	5
Steel chips, crushed	100-150	D28	6	5
Steel turnings	75-150	H38X	6	5
Sulphur, crushed, 1/2 in. and under	50-60	C26S	1	5
Sulphur, 3 in. and under	80-85	D26S	6	5

Sulphur, powdered	50-60	B26SY	1	5
Talc	40-60	B27	1	5
Triple super phosphate	50-55	B27NR	1	5
Tri-sodium phosphate, granulated	60	B27	1	5
Urea	34-42	C27	1	9
Vermiculite, expanded	16	C36W	2	7^b
Vermiculite, ore	80	D26	6	5
Zinc, concentrate residue	75-80	B28	6	5
Zinc oxide, heavy	30-35	A36Z	1	7^b
Zinc oxide, light	10-15	A36WZ	2	7

Food/Feed/Grains/Agricultural Products

Alfalfa meal	17	B36W	2	7^b
Alfalfa pellets	41-43	D27T	1	5^b
Alfalfa seed	45-50	C26	1	5^b
Baking powder	41	A26	1	7^b
Barley	38	B16S	1	7^b

Table 5.1 (Continued)

Food/Feed/Grains/Agricultural Products	Density (lb/ft³)	Properties	Paddle Configuration for Intermediate and Level	
			High Level	Low
Barley, malted	31	B26	1	7[b]
Barley, meal	28	B26	1	7[b]
Beet pulp, dried	15–20	H36T	2	7[b]
Bluegrass seed	11–25	B26	2	7[b]
Bran	16–36	B26SW	2	7[b]
Brewers grain, spent, dry	25–30	C36	1	7[b]
Brewers grits	33	B26	1	7[b]
Buckwheat	40–42	B26	1	5
Buckwheat flour	40	A26	1	5
Buttermilk, dried	30–35	A36PZ	1	7[b]
Cake mix	30	A26YZ	1	7[b]
Carrots, diced, frozen	44	C26	1	5
Cereal flakes	9–12	C26	2	7
Chocolate, powder	40	A27	1	5

Clover seed	45–50	B26	1	5
Cocoa beans	37	C26T	1	9
Cocoa nibs	35	C27	1	9
Cocoa powder	30–35	A36Z	1	9
Coffee, green bean	42	C26T	1	5
Coffee, ground	25	B26	1	7[b]
Coffee, instant	19	A26	2	7[b]
Coffee, roasted bean	23	C16	1	7[b]
Copra, lumpy	22	D26	1	7[b]
Copra cake, lumpy	25–30	D26	1	7[b]
Copra cake, ground	40–45	B26	1	5
Copra, meal	40–45	B26	1	5
Corn cob, meal	35	B26	1	9
Corn, cracked	45	C16	1	5
Corn, flakes	12	C26	2	7
Corn, germ	21	B26	2	7[b]
Corn, grits	42–43	B26	1	5

Table 5.1 (Continued)

Food/Feed/Grains/Agricultural Products	Density (lb/ft^3)	Proper-ties	Paddle Configuration for Intermediate and High Level	Low
Corn, ground	35	B26	1	9
Corn, meal	30–35	B26	1	9
Corn oil cake	25	B26	1	7[b]
Corn, shelled	45	C16S	1	5
Cotton seed, dry, delinted	35	C26	1	9
Cotton seed, dry, not delinted	18–25	C36	2	7[b]
Cotton seed expeller cake	25	B26	1	7[b]
Cotton seed flakes	20–25	C36	1	7[b]
Cotton seed hulls	12	B26W	2	7
Cotton seed meal	35	B26	1	9
Cow peas	45	C26	1	5
Cream, powdered	38	A26KY	1	7[b]
Egg, powder	35	A26	1	7
Farina	44	B26	1	5

Fish meal	35-40	B36	1	7[b]
Fish scrap	40-50	H36	1	5
Flaxseed	45	B16S	1	5
Flaxseed expeller cake	48-50	B26	1	5
Flaxseed, ground	28	B26	1	7[b]
Flax, screenings	27	C26	1	7[b]
Flour	35	A26Y	1	7[b]
Flour, biscuit	26	A26YZ	1	7[b]
Flour, patent	20	A26YZ	2	7
Flour, spring wheat	30-35	A26YZ	1	7[b]
Flour, winter wheat	34	A16YZ	1	7
Flourseed expeller cake	48-50	B26	1	5
Gelatine, granulated	32	C26T	1	7[b]
Germ oil meal	35-40	B26	1	9
Gluten feed	30-35	B26	1	9
Gluten, meal	40-45	B26	1	5
Graham flour	35-40	A26Y	1	7[b]

Table 5.1 (Continued)

Food/Feed/Grain/Agricultural Products	Density (lb/ft^3)	Proper-ties	Paddle Configuration for Intermediate and High Level	Low
Grits	42–43	B26	1	5
Grass seed	10–12	B26SW	2	7
Hay pellets	40–50	D26	1	5
Hominy	40	C26	1	5
Hominy feed	27	B26	1	9
Hominy meal	44	B26	1	5
Hops, spent, dry	35	H36	1	9
Kafir	40–46	C26	1	5
Lactose	48	B26	1	5
Leather scraps	59	H36	1	5
Linseed oil cake	45–50	B26	1	9
Linseed meal	27	B26	1	7[b]
Maize	40–45	B16S	1	5

Malt, dry, ground	22	B26W	2	7[b]
Malt, dry, spent	10	B26	2	7
Malt, dry, whole	27-30	C26S	1	7[b]
Malt meal	36-40	B26	1	9
Malt sprouts	16	C36	2	7[b]
Malt sugar, ground	35	B36	1	9
Malt sugar, unground	30	B36	1	9
Meat meal	35	B36	1	9
Meat scraps, dried	40-45	H36X	1	5
Mids	35	B26	1	9
Milk, dried, flake	5-6	B26KL	2	2
Milk, malted	30-35	A36KLZ	1	9
Milk, whole, powdered	36	A26KLY	1	7
Mill dust	11	A26	2	2
Mill run feed	15-20	B26	2	7
Millett	38-40	B26	1	9
Milo	40-45	B26	1	5

Table 5.1 (Continued)

Food/Feed/Grain/Agricultural Products	Density (lb/ft^3)	Properties	Paddle Configuration for Intermediate and High Level	Low Level
Mixed feeds, heavy	20-28	B36	1	7[b]
Mixed feeds, light	18-22	B26	2	7
Mustard seed	45	B16S	1	5
Oat flour	30-35	A26YZ	1	7
Oat hulls	8	B26	2	2
Oat meal	45-50	B26	1	5
Oat middlings	35-50	C26	1	7[b]
Oats	26	C16S	1	7
Oats, crimped	19-25	B26	2	7
Oats, crushed	25-30	B26	1	7[b]
Oats, groat	46-47	B26	1	5
Oats, rolled	18	C26SW	2	7
Oyster shells, ground	60-65	C27	1	5

Pablum	9	B26	2	2
Peanust, in shell	15–20	D26T	2	7
Peanuts, shelled	35–45	C26	1	9
Pearl starch	38	A26YZ	1	7
Peas, dried	45–50	C16ST	1	5
Peas, frozen	27	C26	1	7
Popcorn, popped	6–10	H36	2	2
Popcorn, shelled	45	C26	1	5
Potato starch	40	A26S	1	5
Pyrethrum flowers, coarse, ground	20	B26	2	7
Pyrethrum spent flowers	32	B26	1	9
Rape seed	48–50	C26	1	5
Rice, bran	20–21	B26	2	7
Rice, clean	30–35	C26	1	9
Rice, grits	42–45	B26	1	5
Rice, rough	32–36	B26	1	9
Rye, bran	15–20	B26	2	7

Table 5.1 (Continued)

Food/Feed/Grain/Agricultural Products	Density (lb/ft³)	Properties	Paddle Configuration for Intermediate and	
			High Level	Low Level
Rye, middlings	42	B26	1	5
Rye, shorts	32–33	B26	1	9
Rye, malted	33	B26	1	9
Rye, meal	33	B26	1	9
Sawdust, dry	10–30	H36	2	7
Sawdust, ground	20	B36	2	7
Semolina	34	B26	1	9
Sesame seed	27	B26	1	7
Shorts	15	C26	2	7
Sorghum seed, milo	47–52	C26	1	5
Soybeans, whole	45–50	C17S	1	5
Soybeans, cake	45	B26	1	5
Soybeans, cracked	30–40	B26S	1	9

Soybeans, meal	40	B26	1	5
Soybean flour	30–35	A26WY	1	7
Soybeans, split	30–40	B26S	1	9
Soybean hulls, unground	6–7	B26	2	–
Soybean millfeed pellets	37–38	C26	1	9
Starch, corn, powder	25–50	A26SWY	2	7[b]
Starch, lump and pelleted	30	C16	1	9
Starch tablets, crystals	40	C26	1	5
Sugar, beet pulp	12–15	H36	2	7
Sugar, brown	40–50	B26	1	5
Sugar, corn	50	B26	1	5
Sugar, dextrose, granulated	50–55	B26KT	1	5
Sugar, dextrose, powdered	50–55	A26KY	1	5
Sugar, granulated	50–55	B26KT	1	5
Sugar, powdered	50–60	A26KY	1	5
Sugar, raw	55–65	B27Z	1	5
Sugar, lactose, granulated	50–55	B26K	1	5

Table 5.1 (Continued)

Food/Feed/Grain/Agricultural Products	Density (lb/ft³)	Proper-ties	Paddle Configuration for Intermediate and Level	
			High	Low
Sugar, pulverized	55	A26	1	5
Sugar beet, granulated	50-55	B26KT	1	5
Sugar beet, powdered	50-55	A26KY	1	5
Sunflower seed	38	C26	1	9
Tea	12	H36	2	2
Timothy seed	36	B26SW	1	9
Tobacco leaves, dry	12-14	H36TX	2	7
Tobacco scraps	15-25	D36W	2	7
Tobacco snuff	30	B36TY	1	9
Tobacco steams	15	H16X	2	7
Wheat	48	C16S	1	5
Wheat, cracked	40-45	B26S	1	5
Wheat germ	28	B26	1	7

Wheat bran	10-15	B26	2	7
Wheat midds	25-30	B26	1	9
Wheat screenings	17-25	B26	2	7
Wood chips	10-30	D36	2	7[b]
Wood flour	16-36	A26WY	2]	7[b]

[a]For alternate high level, use paddle 4.
[b]For alternate intermediate and low level, use paddle 9.

Table 5.2   Material Handling Characteristics Guide (Courtesy of Monitor Manufacturing, Elburn, Illinois)

Property	Grade
Size	
Very fine (minus 100 mesh)	A
Fine (100 mesh to 1/2 in.)	B
Granular (1/8 to 1/2 in.)	C
Lumpy (lumps 1/2 in. and over)	D
Irregular (fibrous, stringy, etc.)	H
Flowability	
Very free flowing (angle of repose less than 30°)	1
Free flowing (less than 30° angle of repose and more than 45°)	2
Sluggish (angle of repose more than 45°)	3
Abrasiveness	
Nonabrasive	6
Mildly abrasive	7
Very abrasive	8
Special Characteristics	
Contaminable, affecting use or saleability	K
Hygroscopic	L
Highly corrosive	N
Mildly corrosive	P
Gives off dust or fumes, harmful to life	R
Contains explosive dust	S
Degradable, affecting use or saleability	T

Table 5.2 (Continued)

Property	Grade
Very light and fluffy	W
Interlocks or mats to resist digging	X
Aerates and becomes fluid	Y
Packs under pressure	Z

As an example, consider pulverized chalk at -100 mesh. From Table 5.1, the properties of the material are designated as A37YZ. From Table 5.2, (A) signifies that the material is very fine, (3) means sluggishly flowable, (7) means mildly abrasive, (Y) means capable of being fluidized, and (Z) indicates capable of packing under pressure. This type of material is best handled by a paddle configuration shown by drawing (6) for high- or intermediate-level control and by drawing (5) for low-level control in Figure 5.9.

6

Displacement-Type Level Controls

INTRODUCTION

Displacer-actuated level control devices are based on Archimedes' principle which states that the resultant pressure of a fluid on a body immersed in it acts vertically upward through the center of gravity of the displaced fluid and is equal to the weight of the fluid displaced. This resultant upward force exerted by the fluid on the body is known as buoyancy. In a displacement-type level sensing system, the immersed body is referred to as a displacer. These systems differ from float-actuated devices in that instead of the float (displacer) floating on the liquid surface, it is supported by arms that allow some small vertical movement as the liquid level changes. The buoyancy force exerted on the displacer can be measured to reflect level position. The displacer-type systems discussed in this chapter operate on a variable displacement principle.

All displacement level measurements are essentially interface measurements, in that the measured variable is the level between two mediums having different specific gravities (e.g., between liquid and gas, liquid and vapor, or liquid and liquid). The magnitude of the displacer travel depends on the interface change and on the specific gravity difference between the upper and lower mediums. For liquid interface service, the displacer is always completely immersed.

ARCHIMEDES' PRINCIPLE

Nearly eighteen hundred years before Newtonian mechanics was devised, Archimedes found that when a body is immersed partially or

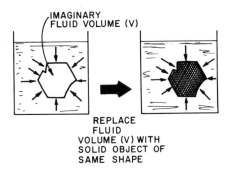

Figure 6.1   This illustrates Archimedes' principle. Hydrostatic
forces act on the surface of an arbitrary volume, V, of liquid within
a liquid. If the liquid volume is replaced by a solid, the same hydrostatic forces are applied.

wholly in a fluid, it is acted upon by an upward or buoyant force
which equals the magnitude of the weight of the fluid displaced by the
body.

To illustrate this point, imagine a volume, V, of a portion of
liquid contained in a vessel (refer to Figure 6.1). The fluid occupying V is at rest because resultant forces from the surrounding liquid
acting on it counterbalance the downward weight of the liquid contained in V. As such, the fluid in V is subject to an upward force
arising from the surrounding liquid which matches the weight of the
liquid within V. If the imaginary volume of liquid is replaced by a
solid having the same configuration (and therefore occupying the
same volume), exactly the same forces act on the boundary of V because the density and other properties of the solid do not affect the
surrounding fluid. Consequently, the solid object in Figure 6.1 is
subject to a force derived from the surrounding fluid which equals
the weight of the liquid it displaces.

The immersed solid body's direction of movement depends on
the resultant of all the acting vertical forces, with buoyant force
being one of the forces. Obviously, if the only external forces in
play are the buoyant force and the weight of the solid, then the solid
body will be accelerated upward if the buoyant force exceeds the
weight. That is, a body that is totally immersed in a fluid will be
accelerated in an upward direction toward the surface if the specific
gravity of the body is less than that of the fluid; and conversely if the
body's specific gravity exceeds that of the fluid, it will sink.

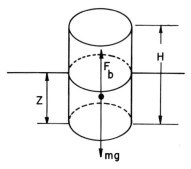

Figure 6.2   Illustrates the forces acting on a cylindrical object immersed in a liquid to a depth Z.

To examine the forces involved in more detail, let us consider a
cylindrical buoy partially immersed in a bath of water and having a
mass m and cross-sectional area A. Figure 6.2 illustrates the system under consideration. The overall height of the object is H and
it is immersed in the liquid to a depth Z. There are two forces acting on the buoy: a downward force derived from its weight (mg), and
an upward buoyant force F_b. The volume of water displaced by the
buoy is AZ. The mass of the displaced water is ρAZ, where ρ is
the density of water, and consequently its weight is ρAZg. A force
balance on the object shown in Figure 6.2 can be written to obtain the
overall resultant force

$$F_R = F_b - mg = \rho AZg - mg \tag{6.1}$$

where F_R is the overall or resultant force which is in the upward direction. The buoy reaches equilibrium (it floats) when buoyancy and
gravitational forces are equal. That is, when $F_R = 0$, the cylindrical object will float at a depth Z_o, accordingly:

$$\rho AZ_og - mg = 0 \tag{6.2a}$$

or, the depth at which it floats is

$$Z_o = \frac{mg}{\rho Ag} \tag{6.2b}$$

Equation (6.2b) can be expressed in a more usable form by replacing the buoy mass, m, by $\rho_b AH$, where ρ_b is the density of the object and H is its overall height:

$$Z_o = \frac{\rho_b AHg}{\rho Ag} = H\frac{\rho_b}{\rho} \tag{6.3}$$

Note that the ratio Z_o/H represents the buoy's volume beneath the fluid's surface and is equal to the relative density of buoy and fluid (ρ_b/ρ), which in this case is the buoy's specific gravity since the fluid is water. As an example, if the buoy's specific gravity is 0.6, then it will float 60%.

If the buoy is released from a nonequilibrium position within the water, then its relative motion can be described as follows:

$$F_R = \rho AZg - \rho AZ_o g = \rho Ag(Z - Z_o) \tag{6.4}$$

The resultant force acting on the buoy is proportional to its displacement $(Z - Z_o)$ from its equilibrium position. If frictional forces are indeed negligible as we have assumed in our derivation, then the buoy will oscillate vertically in simple harmonic motion when released from the nonequilibrium position.

Displacer-actuated level indicators operate on the Archimedian principle by using the buoyant force acting on a partially submerged displacer as a measure of the location of a fluid's interface along the axis of the displacer. The vertical motion of the displacer is normally restrained by some type of elastic member whose motion is directly proportional to the buoyant force, and hence, the fluid level. Figure 6.3 illustrates the basic method behind displacement-type level indicators. The vessel shown in Figure 6.3 has a displacer suspended by a spring scale. The displacer has a diameter of 3.5 in., an overall length of 16 in., and weighs 5 lb. In the vessel to the left, the liquid (water) level is just below the bottom of the displacer. Therefore, the spring scale supports the full weight of the displacer as indicated by the weight scale at 5 lb. If the vessel's water level is increased to the 10-in. level, for example, the displacer becomes partially immersed with the volume of water displaced being 96.2 in.^{3}. With the density of water being 0.036 lb/in^3, the weight of the water displaced is 3.47 lb. From our earlier discussion, the weight of water displaced is equal to the loss in

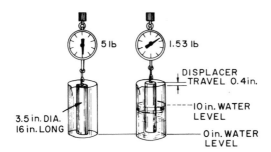

Figure 6.3   Illustrates the operating principle of displacement-type liquid level indicators.

displacer weight. Consequently, the net weight of the displacer is now 1.53 lb., as shown by the scale on the vessel to the right in Figure 6.3

It should also be noted that as the net weight of the displacer has been decreased, the spring scale has lifted the displacer. The amount that the displacer has been lifted is directly proportional to the increase in water level.

TORQUE-TUBE DISPLACERS

The system described by Figure 6.3 can be designed to use a highly accurate spring scale which can be calibrated in terms of level for liquids of known specific gravity. The simple tension-spring design has been altered by a torque-tube and torque-arm design. The torque tube is specified according to relatively tight tolerances with respect to diameter, wall thickness, and length. This provides a desired spring rate for the torque-arm length. The change in the displacer position is slight in comparison with the liquid level change. Consequently, the range of measurement of liquid level in general is greatly increased with this type of device. These systems are generally used for liquid level measurement in closed vessels under pressure.

Figure 6.4 illustrates the basic construction of a torque-tube displacer device. The displacer's success as a level indicator can be attributed to the design of a suitable torque tube which allows the motion of the displacer to be transmitted outside the pressure zone

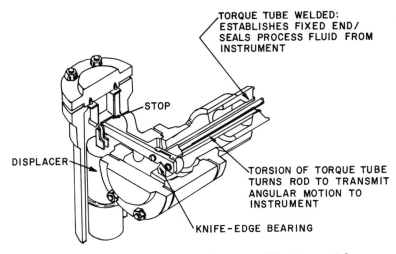

TORQUE TUBE WELDED:
ESTABLISHES FIXED END/
SEALS PROCESS FLUID FROM
INSTRUMENT

STOP

DISPLACER

TORSION OF TORQUE TUBE
TURNS ROD TO TRANSMIT
ANGULAR MOTION TO
INSTRUMENT

KNIFE-EDGE BEARING

Figure 6.4   Shows the basic design features of the torque-tube-displacer level indicator.

without the use of a stuffing box or similar seal. The twist of the torque tube (which is also the rotation of the shaft actuating the level instrument) is typically on the order of 5° [29].

It should be noted that older terminology refer to the displacer as a _float_; however, this is not entirely correct. Displacers are normally weighted to prevent floating and to eliminate backlash (that is, displacers are designed to have specific gravities in excess of the fluid being measured).

These systems are often used with some type of pneumatic or electronic transmitter or controller, as the rotation of the shaft is limited.

MAGNETICALLY COUPLED AND FLEXURE-TUBE DISPLACERS

The operating principle behind a displacer-actuated unit that uses a magnetic coupling is shown in Figure 6.5. Such systems employ a displacer constrained by a spring mechanism that moves a drive magnet enclosed in a protecting tube. The motion of the drive magnet is transmitted to the indicating mechanism by a magnet follower

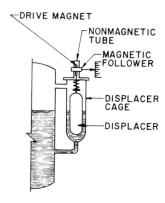

Figure 6.5   Illustrates the basic design and operating principle of
magnetically coupled displacer systems.

installed outside the protecting tube. These systems are generally
mounted in external displacer cages.

   The flexure-tube displacer is an even simpler device and has
previously been described as a float-actuated device becuase of the
similarities in design. Figure 4.1(b) illustrates the major design
features, which consist of an elliptical or cylindrical float mounted
on a short arm connected to the free end of a flexible shaft. The
fixed-in end of the shaft is connected to a mounting flange. A small
segment of the shaft near the mounting flange is flattened; this increases its flexibility. The float's motion is transmitted outside the
float chamber via a rod which extends out through the flexible shaft
or tube. These systems are commonly employed in on-off service
and to actuate directly either an electrical switch or a pneumatic
pilot.

AUTOMATIC TANK GAUGING SYSTEMS

Displacer-actuated devices in general are employed in a variety of
operations. One example is inventory control of liquified gasses
stored under cryogenic or refrigerated conditions. An application
such as this requires special level, as well as temperature gauging
systems. Figure 6.6 shows the details of one manufacturer's design
for this application. The working principle is based on a spring

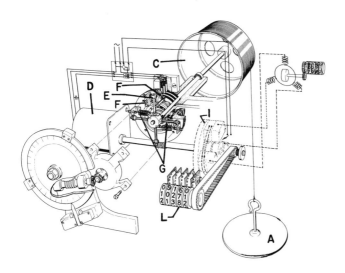

Figure 6.6 Shows key features of one manufacturer's displacer-
actuated level indicator for tank gauging. A displacer (A) is suspended from a flexible measuring cable stored on a grooved measuring drum (C). The drum shaft is coupled to a weighing balance
consisting of a center contact (E), tensioned by two springs (G),
moving between two side contacts (F-F). At equilibrium, the weight
of the displacer, partially immersed in the liquid, balances against
the force of the balance springs. A level rise or fall causes a variation in buoyancy and the weighing-balance center contact moves to
either one of the side contacts. The contacts operate via an inte-
grating circuit a reversible servo motor (D) which turns the measuring drum, thus raising or lowering the displacer until the balance
position is restored. A step transmitter (1) is coupled to the servo
motor shaft which turns 1 revolution per 5-mm level variation,
driving a remote digital indicator. These systems are equipped with
a shaft digitizer (L). (Courtesy of ENRAF-NONIUS Service Corp.,
Bohemia, New York.)

mechanism which functions as follows. A small, solid displacer is
suspended from a flexible but strong measuring cable stored on a
grooved drum. The drum is mounted on a shaft coupled to a weighing
balance which controls a servo motor.

The weight of the displacer partly immersed in the product balances in equilibrium position against the force of the balance springs.
A level variation causes buoyancy variation and changes the balance
position which operates the servo motor, thus turning the drum and
raising or lowering the displacer until the equilibrium position is
restored. The servo motor has a delayed action and averages the
liquid level through an integration circuit, which eliminates the effect
of liquid turbulence. The application of a solid displacer disk with a
relative density higher than that of the product, combined with the
turbulence integration circuit, keeps the wear of the servo system at
a minimum and assures a stable reading—which is particularly important when measuring cryogenic products with a continuously boiling surface [30]. A step-type transmitter coupled to the servo motor
shaft, which turns 1 revolution per 5-mm level variation, drives a
remote digital indicator. As many as three indicators can be connected in parallel to one transmitter. Level gauges connected to a
central selective receiver are equipped with a digitizer.

The level gauge can be equipped with three adjustable alarm
switches. A local digital indicator visible through a window in the
servo motor compartment can be fitted, as well. Each level gauge
is provided with a test circuit. By means of an external push button
or switch it is possible to energize the servo motor, thus hoisting
the displacer. This circuit is intended to carry out repeatability
checks of the level reading and to enable complete retraction of displacer and measuring cable from the tank (e.g., for recalibration
purposes).

Level gauges for cryogenic and refrigerated storage tanks are
composed of two compartments: a gastight drum compartment and
a flameproof servo compartment with terminal box. The drum compartment and servo compartment are separated by an atmospheric
ventilation gap to guarantee safe operation in a Zone 0 area (see
Chapter 2).

With this sort of gauge, only the drum compartment is in direct
contact with the tank atmosphere. This compartment contains a
rotating measuring drum. All other detection parts are housed in
the servo compartment. Drum cover and magnetic cap are sealed by
0-rings. A magnetic coupling transmits the drum rotation to the
servo compartment and vice versa.

The smooth internal shape of the drum compartment is self-
draining and prevents possible deposits from interfering with the
functioning of the instrument. The servo compartment is bolted to

the drum compartment and houses the weighing balance, servo motor, control circuitry, level alarm contacts, transmitter, local indication, etc.

The servo compartment is flameproof and provided with a flanged cover; a separate junction box houses the terminals. The flameproof servo compartment is provided with a heater to prevent winter freezing of the gauge head. The location of the gauge head on the tank top can be freely selected to ensure optimum performance.

MOUNTING CONSIDERATIONS

Proper mounting of the level gauge is essential to safe and accurate measurement with displacement-actuated devices. Maximum measurement accuracy is obtained when the mounting flange on the tank forms a stable reference that is not affected by the deformation of the tank due to vapor pressure, hydrostatic pressure, or roof movements.

Safety of gauging systems encompasses two considerations, namely, flameproofing of terminal boxes and transmitter arrangements, and the safety of the instrument in respect to its use on a tank (i.e., its possible connection in a Zone 0 area). Figure 6.7 illustrates the recommended rooftop mounting arrangement for the level gauge system for liquified gas installations described above. The servo-powered level gauge contains no electrical circuitry whatsoever inside the Zone 0 area. The location of the gauge at the top is strongly recommended in view of safety limitations. Eventual gas leaks usually are quickly diluted by surrounding air before the gas reaches ground level. Location of a gauge at ground level is generally undesirable, as a leak in the gauge or piping connections can result in dangerous gas contamination at ground level, and explosions may follow. Tank-top-mounted gauges are limited in the number of connections and all of them must be sealed to minimize chances for leakage.

Figure 6.8 illustrates the recommended mounting configuration on a cryogenic storage tank. The arrangement consists of a perforated standpipe which is supported by the tank base and passes through both roofs by means of flexible seals. The standpipe is normally fitted with a mounting flange at the top and provides a stable mounting position for the gauging assembly. The pipe used as a still tube or cage for the displacer must be straight and internally free from welding burrs and other obstructions. As shown, a full-bore ball or gate valve is mounted on top of the pipe; this is coupled with a

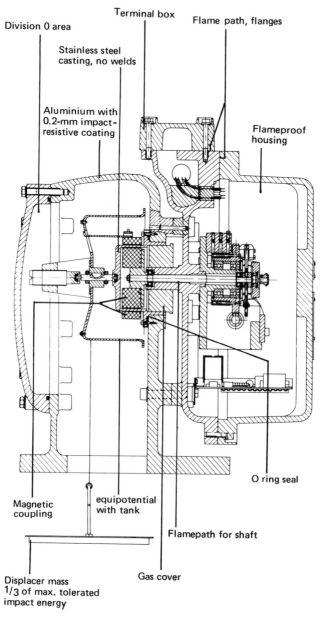

Division 0 area

Stainless steel
casting, no welds

Terminal box

Flame path, flanges

Aluminium with
0.2-mm impact-
resistive coating

Flameproof
housing

O ring seal

Magnetic
coupling

equipotential
with tank

Flamepath for shaft

Displacer mass
1/3 of max. tolerated
impact energy

Gas cover

Figure 6.7   Illustrates key safety design features and proper tank-
top mounting for the displacer-actuated level controller previously
described. (Courtesy of ENRAF-NONIUS Service Corp., Bohemia,
New York.)

126

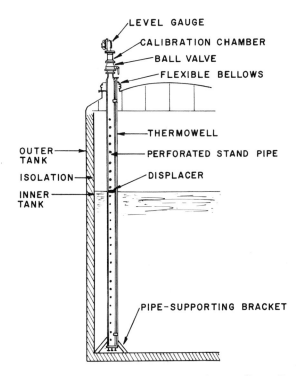

Figure 6.8   Shows proper mounting configuration on a cryogenic
storage tank.

calibration chamber located above the valve. The arrangement
shown in Figure 6.8 allows a complete isolation of the level gauge.
The displacer can be withdrawn from the tank by operating a remote
test circuit. Once the displacer has stopped and is out of the tank,
the valve is closed and the level gauge, measuring cable, and displacer are completely isolated from the tank atmosphere. Inspection
and maintenance of the system is then possible. In existing vessels,
retrofitting with a pipe-still arrangement is not always feasible.
Only heavier displacers can be employed in such cases. It should be
noted, however, that when the gauge is mounted firmly to the tank
shell, it is still less stable and may vary with roof movements. Consequently, the overall accuracy of measurements may be affected by
uncontrollable tank deformations.

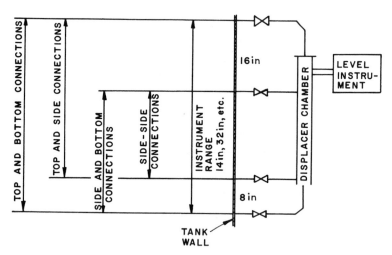

Figure 6.9 Summarizes the possible connection arrangements for external displacer installations.

    There are, of course, many other types of installation configu-
rations, with standard connections being a function of application and
instrumentation range. In principle, a displacer can be designed for
almost any range. The primary requirement is to adjust the size of
the displacer so that the selected range provides the standard rotation; so that, say, for a displacer-torque system, the torque shaft
rotation is standardized. A 14-in. level range, for example, would
use a 3-in. diameter displacer. Aside from the cryogenic applications described above, there are many applications where displacers
are external; consequently the need for a cage for the displacer can
require considerable custom work. Common applications limit selection to available standard lengths of 14, 32, 48, 60, 72, 84, 96,
and 120 in.
    Vessel connection sizes for external installations are typically
1-1/2 in. The lower connection is generally made into the chamber
bottom or side while the upper connection is made into the cage top
or side. Consequently, four combinations are possible. Center-to-
center dimensions of side connections are made the same as the level
instrument range. Figure 6.9 illustrates the possible external displacer cage connections. The arrangement selection depends on
several factors in addition to mechanical considerations. One

consideration is the manometer effect, which causes the weight of
the liquid in the external chamber to balance the column of liquid inside the tank. If the two liquids have different densities brought
about by cooling of the external leg or condensation of light material
in the upper portion of the external chamber, then the external and
internal liquid levels will not be the same. This type of gauging arrangement for a cryogenic application would introduce significant
error to displacer measurements.

The magnitude of measurement error introduced depends on the
location of the connections. If the external chamber starts at the
bottom of the displacer (i.e., top-side or side-side connections), the
error in the indicated level will be introduced at the top of the range
with the lighter material in the chamber. In this case, a less than
full level would be indicated by the displacer fully submerged because of the decreased buoyancy of the lighter liquid. Conversely, if
the external leg starts below the bottom of the displacer (i.e., top-
bottom or side-bottom connections) then the error in the indicated
level is a function of the density difference and the length of the vertical extension of the lower connection. This type of error can be
extremely large.

As a general rule, bottom process-vessel connections should be
avoided if possible. The piping for this type of arrangement tends to
trap water and sediment. If bottom connections must be used, sediment problems and water accumulation can be minimized by extend-
ing the connection several inches into the tank (usually at least 3 or
4 in.). Piping should also be provided with a trap and/or drain.

Top process-vessel connections should also be avoided for vessels which are liquid-filled; this avoids gas trapping. Also, the top
displacer connection becomes inconvenient when maintenance of
internals is needed. Top vessel connections should always be provided with vents.

Finally it should be noted that connections should never be made
into lines that can have flow either to or from the tank (i.e., inlet,
outlet lines, drains, etc.). Velocity head effects from such flows
can introduce enormous errors over the entire instrument range.

Displacer-actuated level controllers are widely employed in a
range of liquid services, including liquid-liquid. For many applications they offer reliable and accurate level detection. The limitations of these systems are generally of an economic nature. Usually
differential-type level instrumentation is more economical when displacer sizes (diameters) exceed 48 in.

NOMENCLATURE

A	area, ft^2
F	force, lb_f
g	gravitational acceleration constant, 32.2 ft/s^2
H	height, ft
m	mass, lb_m
Z	vertical position, ft
ρ	density, lb/ft^3

Subscripts

b	refers to bob or float
o	refers to equilibrium state
R	refers to resultant or total

7

Hydrostatic Head Devices

INTRODUCTION

There are a variety of level detection techniques that are based on the principle of measuring the hydrostatic head. The term head is often expressed in units of pressure or level height. A variety of static and differential pressure techniques is discussed in this chapter. Included are examples of typical control schemes employed with these devices. It should be noted that manometers may also be classified among this group.

PRINCIPLES OF PRESSURE

The term pressure refers to a force per unit area exerted by a fluid on the boundary containing the fluid. Consider the system shown in Figure 7.1 in which we have a cube of water occupying a volume of 1 ft^3. The weight of a cube of water is 62.4 lb$_m$; i.e., the density of the fluid is 62.4 lb$_m$/ft^3 and the fluid exerts a force on the bottom plane equivalent to

$$F = 62.4 \text{ lb}_m/\text{ft}^3 \text{ x } 1 \text{ ft}^3 \text{ x } \frac{g(\text{ft/s}^2)}{g_c[(\text{ft})(\text{lb}_m)/(\text{s}^2)(\text{lb}_f)]} = 62.4 \text{ lb}_f$$

Pressure exerted on the sides of the cube is a function of depth. The pressure exerted on the bottom plane is

$$P = \frac{F}{A} = \frac{62.4 \text{ lb}_f}{1 \text{ ft}^2} = \frac{62.4 \text{ lb}_f}{\text{ft}^2} \tag{7.1}$$

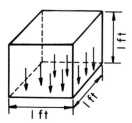

Figure 7.1   Illustrates the physical interpretation of pressure.

Relative and absolute pressures are defined in terms of the
method of measurement. An open-end manometer as shown in Figure 3.2(b), for example, provides a measure of the relative pressure. The open end of the manometer provides the pressure of the
atmosphere as a reference which the process pressure is measured
against. If the process measurement is made against a complete
vacuum (e.g., by sealing off the end of the manometer normally open
to the atmosphere), the measured value is referred to as an absolute
pressure. Absolute pressure is based on that of a complete vacuum;
i.e., it is a fixed reference point which is not a function of location,
temperature, or other factors.

The term <u>barometric pressure</u> refers to a measurement of the
atmospheric pressure using a barometer. A discussion of the principle upon which manometers operate was given in Chapter 3 and is
important to understanding the relationships among pressure measurement definitions.

Gauge pressure is that pressure expressed as a quantity measured from above atmospheric pressure or some other reference
pressure. The relationship between gauge, barometric, and absolute pressures is as follows:

Gauge Pressure + Barometric Pressure
                                    = Absolute Pressure   (7.2)

Figure 7.2 defines the pressure relationships graphically, along
with some other common engineering units.

It should be noted that, for many applications, it is desirable to
reverse the usual direction of measurement and to determine pressures from the barometric pressure down to vacuum. In such cases,
a perfect vacuum becomes the highest pressure on the scale. This
is the same as evacuating the air at the top of a mercury barometer

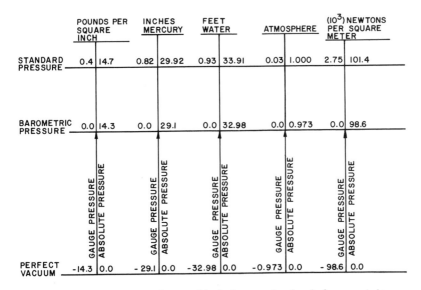

	POUNDS PER SQUARE INCH		INCHES MERCURY		FEET WATER		ATMOSPHERE		(10^3) NEWTONS PER SQUARE METER	
STANDARD PRESSURE	0.4	14.7	0.82	29.92	0.93	33.91	0.03	1.000	2.75	101.4
BAROMETRIC PRESSURE	0.0	14.3	0.0	29.1	0.0	32.98	0.0	0.973	0.0	98.6
PERFECT VACUUM	-14.3	0.0	-29.1	0.0	-32.98	0.0	-0.973	0.0	-98.6	0.0

Figure 7.2   Shows the relationship between standard, barometric, and vacuum pressure at different engineering units.   Standard atmosphere is the pressure obtained in a standard gravitational field and is equivalent to 14.696 lb/in^2 or 760 mm mercury at 0 °C.   Atmospheric pressure is a variable that must be measured with a barometer each time.

and observing the barometer liquid rising up in the barometer as air is removed.   Vacuum systems of pressure measurement are commonly used in apparatis which operate at pressures less than atmospheric.   Pressures which are only slightly below barometric pressure are generally expressed as "drafts" (identical to the vacuum system) and are expressed in inches of water.

Many of the methods discussed in this chapter are based on the principle of measuring the hydrostatic head.   This means that pressures are measured by means of the height of a column of liquid. The term head refers to the weight of liquid above a reference or datum line.   At any point, the liquid's force is exerted equally in all directions and is independent of the volume of fluid or the configuration of its container.   The units of head are those of pressure or level height.   The following relation defines the measurement of pressure due to liquid head, which can be translated to level height above some datum plane:

$$Z = \frac{P}{\rho g}$$ (7.3)

where   P = pressure due to hydrostatic head,

        ρ = density of liquid,

        Z = height of level,

and   g = acceleration of gravity.

Temperature variations may be sufficient to alter liquid density appreciably, which could affect measurement accuracy. This factor must be given careful consideration when automatic level control is utilized.

Pressures greater than atmospheric can be imposed on the surface of a liquid within a closed vessel. This pressure is in addition to the hydrostatic head and must be accounted for with proper instruments to measure differential pressure. The additional pressure exerted by the vapor space above the liquid is imposed on both high- and low-pressure connections of differential pressure instruments and will therefore cancel out mechanically.

PRESSURE GAUGES

Level measurements based on the pressure exerted by the liquid head assume constant density. This means that the instrument must be calibrated for a specific liquid density. Variations in liquid density will result in measurement inaccuracies.

A very common method for measuring liquid level in an open vessel is to connect a pressure gauge below the lowest level to be detected. This level becomes a reference plane and the static pressure detected by the gauge is a measure of the height of the liquid column above the gauge (and hence, of the fluid level). This approach lends itself to a wide range of applications including highly viscous liquids and those containing small concentrations of suspended solids. Proper seals or diaphragms are however necessary for the latter applications.

A seal involves the use of a fluid different from that being measured to fill the measuring system. The filling liquid's free surface
is in direct contact with the measured liquid. It is important that
the two liquids do not mix nor react chemically.

A diaphragm cuts off the liquid in the measuring system from
the liquid being metered. These systems respond to changes in
liquid level via an increase or decrease in the diaphragm's deflection
caused by changes in static pressure exerted upon it by such changes.
A diaphragm transmits with the pressure element through a capillary
tubing that is filled with an inert liquid. The movement of the diaphragm is directly transmitted to the pressure element. When pressure gauges are used for liquid level measurement, they can be calibrated in pressure units, in liquid level units corresponding to the
liquid's specific gravity, or in volumetric units based on the vessel
dimensions. Sometimes it is convenient to calibrate in terms of percentage of maximum level [31].

In measuring liquid level in an open vessel via a pressure gauge,
only one pressure connection is needed. This connection serves to
transmit the sum of the liquid head and atmospheric pressure to the
high-pressure connection of the instrument. The low-pressure connection is thus left open to the atmosphere. Consequently, the overall differential pressure on the measuring device is due to only the
liquid head. As noted above, this type of installation generally requires that the unit be mounted at approximately the same elevation
as the reference level. Elevation adjustments are sometimes provided to correct for up to 5% of the range [20].

Figure 7.3 shows a typical bellows differential pressure (D/P)
indicator designed to measure differential pressures, particularly of
flow and liquid level. The basic D/P indicator consists of a high-
and low-pressure bellows, both liquid-filled and connected to a center plate. The bellows are enclosed in the high- and low-pressure
end housings which in turn are bolted to a center plate. When two
different pressures are applied to the high- and low-pressure connections, the high-pressure bellows contract, forcing the fill fluid
through the center plate into the low-pressure bellows, thereby causing it to expand.

The resulting motion of the low-pressure bellows can be transmitted through a temperature-compensated linkage to the instrument
output shaft. This linear motion is simultaneously converted to a rotary motion of the output shaft. For the unit shown in Figure 7.3, the
output shaft is part of a low friction, O-ring-sealed shaft-and-bearing

Figure 7.3   Typical pressure gauge used for flow measurement and level detection applications. D/P units such as this are widely used in industrial process plants, power plants, and pollution control facilities as well as for cryogenic liquid storage tank level indication. (Courtesy of Meriam Instrument, Cleveland, Ohio.)

assembly which transmits the rotary motion to the external area of the bellows unit and into the indicator case. The torque required by the shaft-and-bearing assembly is less than that needed by a torque tube which could also be employed. The rotary motion of the output shaft is transmitted to the indicator pointer through conventional linkage-and-movement mechanisms.

These units are normally equipped with a dampening valve in the center plate through which the fill fluid must pass. The valve can be field-adjusted to achieve the desired dampening effect.

DIFFERENTIAL PRESSURE TRANSMITTERS

Differential pressure (D/P) transmitters are often referred to as the "work horse" of process measurements. They have been applied to

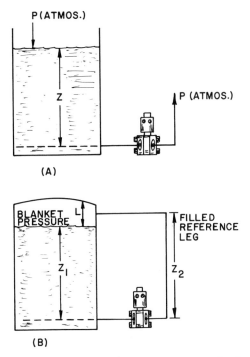

Figure 7.4   (A) D/P transmitter applied to measuring liquid level in an open vessel by measuring the static head (Z). (B) D/P transmitter applied to a closed vessel.

a multitude of process measuring problems, including measurement of flow, differential pressure, specific gravity, absolute pressure, as well as level.

As noted earlier, a liquid in an open tank exerts a hydrostatic head of pressure at the bottom of the vessel that is proportional to its density and its depth. This is expressed by Equation (7.3). Sensing this pressure with a gauge pressure transmitter provides an accurate indication of the liquid depth. Today's direct sensing transmitters have accuracies to within $\pm 0.25\%$ [32]. Figure 7.4(A) illustrates level measurement application in an open tank.

In a closed or pressurized vessel, the pressure at the bottom of the tank consists of the hydrostatic pressure plus the pressure in the vapor space of the tank above the liquid. This can be expressed as

$$P = (Z \times G \times 62.4) + P_R + \rho_v L \qquad\qquad (7.4)$$

where P_R = pressure in the tank vapor space (lb_f/ft^2),

           Z = height of liquid (ft),

           G = liquid-specific gravity referenced to water,

           ρ_v = is the density of the vapor/gas above the liquid (lb_m/ft^3),

and L = height of the vapor space (ft).

To sense this pressure with a D/P transmitter, the high-pressure port of the unit is connected to the tank bottom and the low-pressure port is connected to the vapor space. This arrangement, illustrated in Figure 7.4(B) provides an accurate indication of the liquid depth by eliminating the effect of tank pressure above the liquid. It should be noted that variations in the vapor space can result in erroneous level measurements. By filling the reference leg with a seal liquid, this problem can be minimized.

The difference between the process fluid level Z and that of the sealed fluid Z_2 in Figure 7.4(B) is proportional to the tank level. In this example the minimum differential exists when the vessel is full, and the maximum occurs when Z_1 is zero. In order to read level directly, the transmitter output must be reversed so that the maximum output takes place at the minimum input.

In comparison to switch-like float or ultrasonic detector instruments (to be discussed later) which can be used under certain conditions only to control liquid level to a mechanically preset point, D/P transmitters permit both continuous measurement and control, at the site or remotely. In addition, through the use of supplementary on-site or remote controllers, these devices offer almost infinitely adjustable level control at any point within the vessel, with little or no alteration in the mechanical installation [33]. This enables programmed leveling, high or low limit variations, automatic fluid density adjustments, and so forth.

Most D/P transmitters utilize the movement of a precision pressure-measuring diaphragm to induce a noncontracting differential magnetic coupling which is then transformed by solid-state electronic circuitry into an output signal (typically 4-20 mA dc) [33]. No levers, linkages, or mechanical servos are used with these systems.

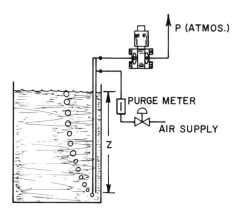

Figure 7.5   Illustrates the use of a bubbler line with a D/P transmitter. This arrangement minimizes line plugging from suspended
solids in the process liquid.

The signal can be fed directly to recorders, indicators, and/or controllers. There is a wide range of installation arrangements suited
for specific applications, of which the most widely used are described below, along with recommendations for their use.

   For process liquids or slurries containing particulates, the arrangement shown in Figure 7.5 is often employed. This consists of
a regulated air supply which is fed through a bubbler pipe which is
mounted in the vessel. The backpressure from the bubbler pipe
matches the liquid's hydrostatic head and is connected to the high
side of the D/P transmitter. The low-pressure side is generally
vented to the atmosphere so that the transmitter's output is proportional to the static head. The gas is usually air but can be any gas
that is compatible with the process liquid. The gas pressure is the
parameter measured by the pressure transmitter and is taken as an
indication of the liquid level. This arrangement is especially suited
for high-temperature installations, dangerous fluids, and slurries.
Only the dip-tube material requires special consideration when the
liquid is corrosive. Generally, clearance for sediment must be provided between the dip-tube and tank bottom. In some applications,
fouling of the dip-tube outlet from solids or other contaminants is a
concern. In these cases, provision can be made for periodic

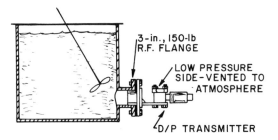

Figure 7.6   Illustrates the use of a tank level transmitter.

high-pressure blowdown. These arrangements must be provided with
a suitable transmitter bypass to prevent overload damage.

Many storage tank applications preclude the use of conventional
mounting configurations. Examples include highly viscous fluids,
fluids that require heat tracing and jacketed vessels, tanks requiring
agitation, and highly corrosive fluids. For these situations a tank
level transmitter, as illustrated in Figure 7.6, is often employed.
This arrangement consists of a D/P transmitter having a flanged
connection on the high-pressure side.

The process pressure is exerted on the isolation diaphragm
while the pressure is transmitted through a capillary tubing filled
with silicone oil to the high side of the transducer. The low side is
generally vented to the atmosphere, although it sometimes is connected to the vapor space of a closed vessel. The transmitter can be
mounted to a mating process flange in the vessel having an extension
length established by the thickness of the tank wall and/or the throat
of the tank fitting. Some applications do not require this extension.
It is only necessary to isolate the diaphragm when extreme conditions
exist (i.e., solids, very high viscosities, etc.).

The arrangement illustrated in Figure 7.6 permits measurement
of the static head at a point that is nearly flush with the vessel wall.
This has the advantage of preventing plugging or process buildup in
front of the sensor unit. The main disadvantage in making the transmitter an integral part of the vessel is that the tank must be taken
out of service and drained in order to maintain and recalibrate the
transmitter.

Table 7.1 summarizes other pressure transmitter installations
commonly used. The first arrangement consists of a small air
chamber at the bottom of the tank which is connected by small-bore

Table 7.1   Various D/P Transmitter Installation Arrangements

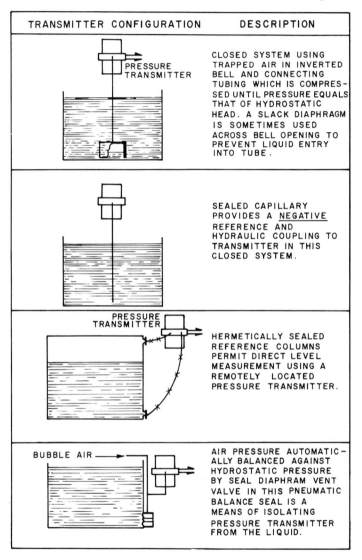

TRANSMITTER CONFIGURATION	DESCRIPTION
PRESSURE TRANSMITTER	CLOSED SYSTEM USING TRAPPED AIR IN INVERTED BELL AND CONNECTING TUBING WHICH IS COMPRESSED UNTIL PRESSURE EQUALS THAT OF HYDROSTATIC HEAD. A SLACK DIAPHRAGM IS SOMETIMES USED ACROSS BELL OPENING TO PREVENT LIQUID ENTRY INTO TUBE.
	SEALED CAPILLARY PROVIDES A <u>NEGATIVE</u> REFERENCE AND HYDRAULIC COUPLING TO TRANSMITTER IN THIS CLOSED SYSTEM.
PRESSURE TRANSMITTER	HERMETICALLY SEALED REFERENCE COLUMNS PERMIT DIRECT LEVEL MEASUREMENT USING A REMOTELY LOCATED PRESSURE TRANSMITTER.
BUBBLE AIR	AIR PRESSURE AUTOMATICALLY BALANCED AGAINST HYDROSTATIC PRESSURE BY SEAL DIAPHRAM VENT VALVE IN THIS PNEUMATIC BALANCE SEAL IS A MEANS OF ISOLATING PRESSURE TRANSMITTER FROM THE LIQUID.

tubing to a pressure transmitter. A flexible diaphragm across the air-chamber opening can be used to seal the air. The liquid hydrostatic pressure compresses the air in this closed system until the air pressure matches the liquid pressure. The transmitter measures this air pressure, thus providing an indication of the liquid level. This type of installation is well suited for situations where purging is not acceptable or when compressed air is unavailable. It is most often used in open vessels, top-entry installations (including buried tanks and free streams), tidewaters, and tailraces. This closed-chamber design has also found limited application in pressurized tanks under limited pressure conditions. When using corrosive liquids, attention must be given to proper selection of the air-chamber and tubing materials. Distances between the tank and display instrumentation are limited typically to 200 ft [33]. These systems are referred to as linear variable differential transducers (LVDT). It should be noted that a major source of measurement error with these systems is piping leaks.

The second arrangement shows the air chamber connected to the transmitter by a vacuum-filled liquid capillary tube, which creates a negative reference column. If the tank were empty, this arrangement would produce a negative pressure on the transmitter corresponding to one atmosphere, typically. The liquid hydrostatic pressure reduces this negative pressure, with a corresponding reduction in negative pressure transmitter indication proportional to the liquid level. The empty tank situation produces the maximum negative transmitter indication while a full tank produces a zero transmitter indication. This arrangement is limited in range. It can only be used in open or vented tanks. Also, distances between the tank and transmitter are limited.

For the third system illustrated in Table 7.1, inert, liquid-filled capillary tubes between the transmitter and isolation seals mounted on the tank pressure taps allow the removal of the transmitter itself from the vessel vicinity to more convenient or safe locations. It should be noted that remote sensing systems such as this are subject to measuring errors because of environmental temperature variations. In general, however, distances between readout instruments and tanks are unlimited.

The last arrangement shows an alternative method of isolating the transmitter from the liquid. This is done through the use of a pneumatic balance seal. The hydrostatic head against an isolating diaphragm is maintained in balance by air pressure through a venting valve. The air pressure, which matches the hydrostatic head, is measured by a pressure transmitter as in the bubbler system (Figure 7.5). Again, the distances between tank and readout are practically unlimited. References 34-36 should be consulted for further discussions on installation arrangements.

LEVEL INDICATORS AND CONTROLLERS FOR D/P TRANSMITTERS

Indicating and control instrumentation for the systems described above can range from relatively simple circuitry to fairly complex designs. A brief discussion of some typical indicating and control arrangements summarized in Table 7.2 can help illustrate this point as well as provide some background information useful to system selection.

For each of the level sensing systems described above, the output consists of a two-wire pressure transmitter signal (typically 4-20 mA dc). Such a signal can operate indicators and/or recorders and controllers. The complexity of these instruments is dictated by the requirements of the particular application.

The first system shown in Table 7.2 is representative of some of the simplest level indicators. It consists of a 4- to 20-mA panel meter scaled from 0 to 100%. The indicator can be mounted directly on the transmitter or in a remote location. The meter forms a self-contained indicating system. Digital panel meters are also commonly employed. The accuracy of indication by panel meters is typically limited to about 2% of full scale; however, digital readout can provide a system accuracy of about .5% of full scale [37,38].

The second schematic drawing in Table 7.2 is a representative complex type of level indicator typically used for a bubbler installation (Figure 7.5). The transmitter is shown combined with various digital readouts. A unit such as this can be supplied by the manufacturer as a self-contained unit ready for bulkhead mounting. These systems can be equipped with remote control room readout of liquid level, set point lights, and alarms.

Controllers also vary in complexity. The third schematic drawing shows a simple meter relay system to provide level control action. Adjustable pointers on the meter incorporate electrical contacts which actuate electrical relays in the meter. This provides a means for activating tank inlet and outlet valves, pumps, etc., for level control as well as high and low level alarms. A single-meter

Table 7.2 Schematic Drawings of Various Liquid Level Indicators
and Controllers

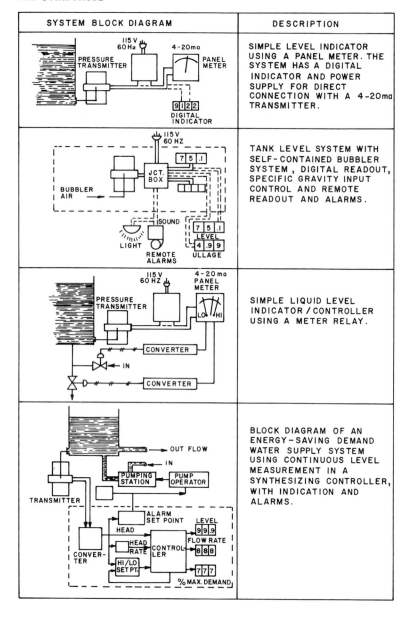

SYSTEM BLOCK DIAGRAM	DESCRIPTION
	SIMPLE LEVEL INDICATOR USING A PANEL METER. THE SYSTEM HAS A DIGITAL INDICATOR AND POWER SUPPLY FOR DIRECT CONNECTION WITH A 4-20ma TRANSMITTER.
	TANK LEVEL SYSTEM WITH SELF-CONTAINED BUBBLER SYSTEM, DIGITAL READOUT, SPECIFIC GRAVITY INPUT CONTROL AND REMOTE READOUT AND ALARMS.
	SIMPLE LIQUID LEVEL INDICATOR/CONTROLLER USING A METER RELAY.
	BLOCK DIAGRAM OF AN ENERGY-SAVING DEMAND WATER SUPPLY SYSTEM USING CONTINUOUS LEVEL MEASUREMENT IN A SYNTHESIZING CONTROLLER, WITH INDICATION AND ALARMS.

relay generally can be used to control both upper and lower level limits to within 3% of full scale [33].

The last drawing illustrates that even more complex control functions are possible through the use of electronic controllers. The system shown represents a demand water supply system. The energy consumption needed to deliver water in a demand supply system can be greatly reduced by employing continuous liquid level measurement to control pumping operations. The hydrostatic head signal from the D/P transmitter serves as the input to a controller, which uses the measured head and rate of change of head to synthesize an optimal-energy-use pump control signal. Such a system can be equipped with various level readouts to indicate flow rates, percentages of maximum demand, set points, etc., along with suitable alarms and signal lights on all process parameters.

MISCELLANEOUS PRESSURE TECHNIQUES

There are a variety of other commercial devices that rely on pressure sensing as an indication of liquid level. One class of sensing instruments uses a pressure-duplicating method. This consists of a level transmitter which converts the pressure of the liquid head into an air signal, which is transmitted to a pressure-measuring instrument as receiver. The transmitter for these devices is mounted on the bottom of the tank and the liquid head exerts a downward force over a diaphragm. An air supply (normally 3 to 5 lb/in^2 above the highest pressure exerted by the liquid head) is connected with the transmitter and an air line connects the transmitter with the receiver. The downward force exerted by the liquid in the tank is counteracted by an upward force from the air pressure which is admitted against the lower side of the diaphragm. As the tank's liquid level rises, the diaphragm undergoes further deflection. This motion is then transmitted through a contact button to a baffle arrangement which moves closer to an air-bleed nozzle. This action restricts the air flow through the bleed nozzle to the atmosphere, causing the air pressure in the diaphragm chamber to increase. The air pressure in the diaphragm chamber becomes equal to the pressure of the liquid head on the diaphragm. The receiver can therefore be calibrated in units of liquid level.

These systems closely resemble air-bubbler arrangements previously described. The primary difference is that the air is not permitted to escape through the liquid but is passed through a bleed nozzle.

These designs can be employed for closed tanks under pressure when used with a pressure differential receiver. A conventional arrangement would involve connecting the high-pressure side with the bottom of the tank and the low-pressure side to another transmitter positioned on the top of the tank. This prevents the possibility of condensing vapors from developing in the low-pressure lines.

For tanks to be measured under vacuum, the air from the diaphragm chamber cannot exhaust to the atmosphere, as the pressure against the bottom of the diaphragm would be too great and balance could not be achieved. In such cases it is necessary to bleed the diaphragm chamber to a vacuum chamber. It is also acceptable to use the vapor space above the liquid as the vacuum chamber.

Finally, there are a whole range of instruments for measuring fluid flow by differential pressure methods that can be applied to level detection in tanks under vacuum or pressure. These devices include orifice meters, flow nozzles, Dall flow tubes, manometers, and others. In applying these instruments to pressure or vacuum tanks, the instruments will provide reverse readings. That is, while the instrument reads zero flow when used as a flowmeter, it will read a maximum level when used for metering liquid level. The working principles behind these instruments are discussed in an earlier volume [39], which the reader can refer to for specific applications.

NOMENCLATURE

A area, ft^2

F force, lb_f

G liquid-specific gravity, referenced to water

g gravitation constant, $32.2 \ ft/s^2$

L height, ft

P liquid pressure, lb/in^2

P_R vapor pressure, lb/in^2

Z liquid height, ft

ρ liquid density, lb/ft^3

ρ_v vapor density, lb/ft^3

8

Electronic Level Sensing Devices

INTRODUCTION

Electronic level sensing techniques are based upon principles of capacitance, conductance, and resistance. Basic operation principles can be applied to level-detecting instrumentation applicable to liquid and bulk solids services. In general, these devices may be considered to be passive systems in that they involve the use of sensors without moving parts. Choice of measuring technique depends on the medium being measured, operating conditions, vessel configuration, and the type and degree of desired performance (i.e., high/low control, high/low alarms, continuous indication and/or recording, digital or analog output with control). Almost all electronic systems are three-component systems, consisting of sensors, amplifiers, and readout devices.

CAPACITIVE TRANSDUCERS AND PROBES

Figure 8.1 schematically represents a capacitive transducer; this is a plate-type capacitor whose capacitance (measured in picofarads) is given by the following formula:

$$C = 0.225\epsilon\frac{A}{d} \qquad (8.1)$$

where   A = overlapping area (in^2),

d = distance between plates (in.),

and ϵ = dielectric constant.

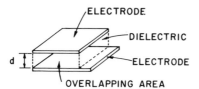

Figure 8.1   Illustrates a capacitive transducer.

A plate arrangement can be used to detect changes in the distance d through variations in capacitance. Also, a change in capacitance may be measured through a change in the overlapping area A that results from the relative movement of the plates in a lateral direction or a change in the dielectric constant of the medium between the plates. Bridge circuits (to be described later) can be used to measure the capacitance. A capacitor's output impedance is given by the following general formula:

$$Z' = \frac{1}{2\pi f C}$$
(8.2)

where Z' = output impedance (Ω),

f = frequency (Hz),

and C = capacitance (f).

The capacitance of a parallel-plate unit is dependent on the plate separation and the dielectric material. In a simple, capacitance liquid level gauge, the liquid would have access to the space between the two parallel plates. Since the dielectric constant of the liquid differs from that of its vapor or air, the capacitance of the unit is a function of liquid height. The dielectric constant ϵ of the medium between the plates is a constant value and differs for various materials. Table 8.1 gives values of ϵ for different materials.

A parallel-plate configuration is inconvenient for tank gauging. Conventional variable capacitor level gauges consist of a cylindrical probe mounted in the tank. The probe, which is isolated from the metallic tank walls, functions as one electrode while the vessel can serve as the other. Figure 8.2 illustrates typical applications of this type of capacitor probe. The level control probes consist of active and inactive lengths and NPT connections to the vessel. The active length of each probe is the portion sensitive to the material

Table 8.1   Dielectric Constants for Various Materials

Material	Dielectric Constant	Comment
Air	1	
Oil	2	
Acids	12–40	
Water	50–80	
Vacuum	1.0	
Sand	3–6	dependent on moisture
Grain	3–4	dependent on moisture

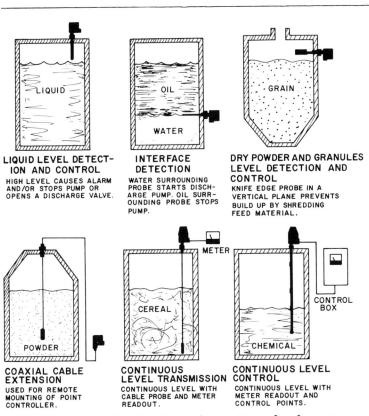

**LIQUID LEVEL DETECT-
ION AND CONTROL**
HIGH LEVEL CAUSES ALARM
AND/OR STOPS PUMP OR
OPENS A DISCHARGE VALVE.

**INTERFACE
DETECTION**
WATER SURROUNDING
PROBE STARTS DISCH-
ARGE PUMP. OIL SURR-
OUNDING PROBE STOPS
PUMP.

**DRY POWDER AND GRANULES
LEVEL DETECTION AND
CONTROL**
KNIFE EDGE PROBE IN A
VERTICAL PLANE PREVENTS
BUILD UP BY SHREDDING
FEED MATERIAL.

**COAXIAL CABLE
EXTENSION**
USED FOR REMOTE
MOUNTING OF POINT
CONTROLLER.

**CONTINUOUS
LEVEL TRANSMISSION**
CONTINUOUS LEVEL WITH
CABLE PROBE AND METER
READOUT.

**CONTINUOUS LEVEL
CONTROL**
CONTINUOUS LEVEL WITH
METER READOUT AND
CONTROL POINTS.

Figure 8.2   Typical applications of capacitance level gauges.

being measured. The inactive segment has the dual purpose of preventing deposits near the hub from causing false operation and leads
the active portion away from the side or top of the vessel. Liquid or
granular materials approaching or submerging a portion of the sensing probe result in a change in capacitance. This change is detected
by an electronic circuit and can be translated into an on-off relay
actuation for control or alarm application. Special considerations
must be given to minimizing the effect of buildup or coating of the
controlled material on the sensor.

For tanks fabricated from insulating material, a suitable second
electrode must be provided. Again, changes in capacitance between
the probe and second electrode are caused by changes in the dielectric value of the media.

Since all capacitance systems rely on a change in capacitance in
order to switch a relay (point control, high or low) or simply to indicate level (i.e., continuous measurement), it is necessary to determine how large a change can be expected. It is reasonable to assume
that only the dielectric constant is important in effecting a change in
capacitance (the discussion here is limited to a fixed probe permanently mounted in a tank, or probe with ground tube, etc.). As al-
ready noted, when a probe is installed into a metal tank, a capacitor
is formed (the tank wall is one plate of the capacitor, the probe wall
becoming the other). If a high-frequency source is connected to the
capacitor, a small current will flow. The amount of current flow is
dependent upon the system's capacity. The capacitive reactance
(i.e., the resistance to current flow) can be estimated from Equation (8.2).

Standard point controllers typically operate at a frequency of
100 kH$_Z$ (continuous operation is at 33 or 100 kH$_Z$) [20]. Because
constant frequency is used, reactance varies only with a change in
the dielectric constant. All other factors are constants which depend
on the particular capacitance controller, electronic insert, probe
configuration, and probe location. Obviously, as the dielectric constant increases the current flowing between the probe and tank wall
or probe and second electrode also increases.

From Table 8.1, note that the dielectric constant of a vacuum
is 1. For practical purposes, air also has a dielectric constant
ϵ of 1 and all other materials have values greater than 1.

The change in capacitance is also proportional to a change in
dielectric constant, as follows:

$$\Delta C = C_a' (\hat{\epsilon} - 1) \qquad\qquad (8.3)$$

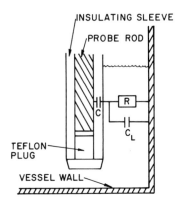

Figure 8.3   Capacitance system used in conductive liquid service.

where $\hat{\epsilon}$ is the relative dielectric constant dielectric value of the material covering the probe and $C_a^{'}$ is the capacity of the probe in air (pf/ft). $C_a^{'}$ is sometimes thought of as the displacement sensitivity of the system. In a general definition, it is the differential of Equation (8.1)

$$C_a^{'} = \frac{\partial C}{\partial d} = -\frac{0.225 \epsilon A}{d^2} \qquad (8.4)$$

Often in liquid service, the liquid displays significant electrical conductivity properties. Examples of conductive liquids are water, acids, bases, etc. These liquids have dielectric constants of approximately 19 and up. For these cases, the change in capacitance is caused by changing the geometrical conditions of the capacitor. Figure 8.3 illustrates this type of system. The ground or reference plate is moved through the liquid directly on to the outer diameter of the insulating tube. This provides a higher capacitance over a small distance than before with a larger distance (between the probe rod and vessel wall).

In addition to straight level detection of liquids and granular solids, capacitance probes can also be used for interface detection in liquid-liquid service. Figure 8.4 illustrates one manufacturer's design. The unit consists of a probe built into a float which senses the surface of the liquid. If the sensor is in water, for example, a high capacitance is registered (water has a relatively high dielectric

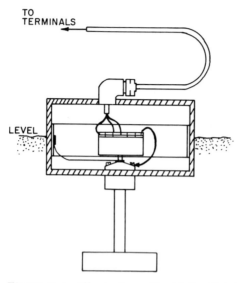

Figure 8.4 Illustrates a liquid-liquid interface detector based on
the capacitance method. A significant dielectric value must exist
between the two fluids in order for this system to function properly.
(Courtesy of Endress & Hauser, Inc., Greenwood, Indiana.)

constant). If, however, a different liquid such as oil (with a lower
dielectric value) floats on top of the water, the probe will register a
lower capacitance. In the electronic insert, the capacitance change
is transformed into a direct current which is passed on to control a
relay switch.

INSTALLATION PRECAUTIONS AND TROUBLESHOOTING

To measure changes in capacitance accurately, the proper probe
configuration must be selected, based on the particular application.
A rod probe is generally considered adequate for use in point level
detection in nonagitated liquids for depths up to about 3 m.

   Liquids having low dielectric constants (e.g., liquid gas stored
in a large-diameter tank), the liquid level changes can result in only
very small capacitance changes. In such cases, a suitable ground
tube should be employed to serve as the second plate of the

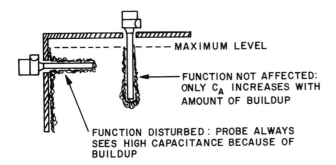

Figure 8.5   Illustrates probes with conductive buildup.

capacitor. This arrangement provides a larger initial capacitance, which in turn results in larger capacitance changes.

For high or low level detection, horizontally mounted probes generally provide a higher indicating accuracy as they are covered or uncovered along the full length of the rod.

Where depths exceed about 3 m, cable probes must be employed. These are suspended from the tank roof and anchored to the tank floor either by a weight attached to the bottom of the probe or by a permanent attachment which prevents the probe from swaying in the tank. Probe movement or cable swaying caused by filling operations (or, e.g., in tanker operations by ship movement) can cause the probe to move closer to the tank wall, thus increasing the capacitance and causing erroneous measurements. The majority of cable probes are limited to depths of roughly 60 ft.

Most commercial units are insulated with heat-resistant materials, such as Teflon. These probes can withstand operating pressures of up to 1500 lb/in^2 and temperatures as high as 400°F [40]. Also, the probe-sensing voltage is sufficiently low so as to classify these systems as intrinsically safe. Amplifiers must, however, be mounted in nonhazardous areas or in explosionproof enclosures.

It should be noted that for continuous analog measurement, it is important that the dielectric constant change over the full length of the probe be linear and gradual. Severe or abrupt changes in the tank environment such as changes in humidity, temperature, or concentration can cause nonlinearity in output signals resulting in significant measurement error.

Conductance probes are subject to a form of fouling known as conductive buildup. Figure 8.5 illustrates this phenomenon.

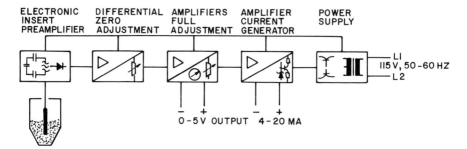

Figure 8.6   Block schematic of capacitance-level gauging system. (Courtesy of Endress & Hauser, Inc., Greenwood, Indiana).

Material buildup on probes will cause the capacity of the unit to in-crease in air. To a degree, this can be compensated for by suitable adjustments provided by the manufacturer.

A block schematic of a capacitance probe and level control sys-tem is given in Figure 8.6. By way of review, the probe and metal tank wall form a capacitor whose capacitance varies with the level. An electronic insert in the probe converts the capacitance measure-ment to a proportional dc output current. This analog signal is fed to an appropriate measuring circuit. These systems are designed with failsafe considerations as illustrated by the sensitivity plots given in Figure 8.7. In Figure 8.7(A) the output relay will de-ener-gize for a high-level condition (i.e., this represents the maximum failsafe condition). If the input power to the controller fails, the output relay will de-energize immediately. Other types of failures within the controller and probe also will cause the relay to de-ener-gize. Note that not all possible failures within the controller and probe cause the relay to drop out. Examples include loose connec-tions to the housing guard of the unit and broken-off probes. Fig-ure 8.7(B) shows that the output relay is de-energized for a low-level condition. Here, conditions are just reverse to that of maxi-mum failsafe.

CAPACITANCE CONTROLLER CALIBRATION PROCEDURES

Recommended calibration procedures for continuous capacitance controllers in liquid service depend on the conductive properties

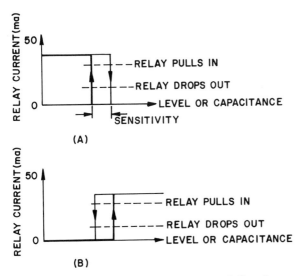

Figure 8.7   (A) Illustrates maximum fail-safe condition for high
level control; (B) Maximum fail-safe condition for low level control.
Note that systems having dead-band function increase the hysteresis
shown here. The distance between the relay being in an energized
and deenergized state will expand to the extent of the adjustment on
the system's potentiometer.

of the liquid and operating limitations of the process vessel. There
are three calibration procedures that can be followed. These are

1.   A full controller calibration where the tank can be re-
     moved from service [Mode (1)]

2.   When the tank cannot be removed from service, but the
     dielectric constant is known [Mode (2)]

3.   When the tank cannot be removed from service and the
     dielectric constant is not known [Mode (3)]

For nonconductive liquids, these procedures are as follows:

Mode (1) This is a full controller calibration in which the process vessel can be removed from service. The capacitance controller
can be calibrated by first emptying the tank and adjusting the system
controller to a "zero" setting. Second, the tank should be filled to

the desired 100% level and the capacitance controller set to indicate that the tank is full. After this initial calibration the following steps are performed:

The electronic insert from the probe head should be disconnected.

The capacitance bridge between the probe and housing guard should be connected.

The capacitance reading should be recorded for the zero setting.

The capacitance reading should be recorded for the full setting.

The electronic insert to the probe should be reconnected.

If the controller is replaced or settings are accidentally changed, the system can be readily recalibrated without emptying the tank. This can be accomplished by using a capacitance decade box or an arrangement of capacitors having the same capacitance values as recorded during the initial installation, to simulate the probe under empty and full conditions.

Mode (2) This method is employed if the dielectric constant is known but if the tank cannot be emptied. The initial and changed capacitances can be computed by the calculation procedures outlined in Figures 8.8 through 8.10. External capacitors can then be used to establish zero and full settings. Figure 8.8 provides the following formulas for computing the change in capacitance for probe mountings in a cylindrical tank:

For top-mounted probes in vertical tanks

$$\Delta C_E = k \Delta C_C \tag{8.5}$$

where ΔC_E, ΔC_C are the capacitance changes for excentric and centric probe mountings, respectively and k is the correction factor for excentrical mounting of a probe (k values can be obtained from the graph provided in Figure 8.8).

For top-mounted probes in horizontal tanks

$$C = 24.14 \frac{\epsilon_r - 1}{\ell g \ D/g} h_x \ 10^{-12} \tag{8.6}$$

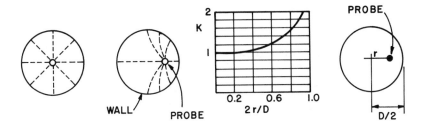

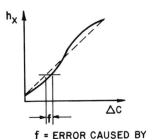

$$\Delta C_E = K \Delta C_C$$

CORRECTION FACTOR FOR EXCENTRICAL MOUNTING

ELECTRICAL FIELD IN A CYLINDRICAL
HORIZONTAL TANK

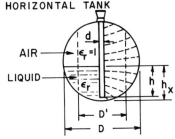

CALCULATION FOR CAPACITANCE:

$$C = 24.14 \frac{\epsilon_r - 1}{\lg D/g} h_x \times 10^{-12} \text{ (FARADS)}$$

NOTE: FOR LARGE TANKS REPLACE D BY
D' = 0.6 D

f = ERROR CAUSED BY
INHOMOGENITY OF
ELECTRICAL FIELD

Figure 8.8   Capacitance change formulas and correction factors for
vertical probe mountings in cylindrical vertical and horizontal tanks.
(Courtesy of Endress & Hauser, Inc., Greenwood, Indiana).

where   D = tank diameter,

   ℓ = probe length,

   C = capacitance (electrical capacitance outgoing from
      formulas for vertical tank,

and h_x = liquid height.

Figure 8.9 provides a nomograph for this equation for quick
calculations.

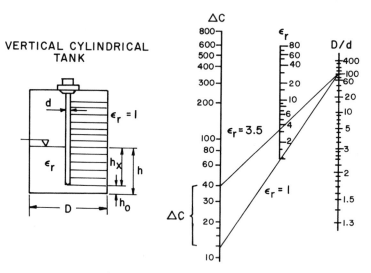

$$\Delta C_{CENTRIC} = 24.16 \, \frac{\epsilon_r - 1}{\lg D/g} \, h_x \times 10^{-12} \text{ (FARADS)}$$

Figure 8.9   Nomograph for Equation 8.6. This is the fundamental expression for capacitance changes of a top-mounted probe (center) in a vertical cylindrical tank. (Courtesy of Endress & Hauser, Inc., Greenwood, Indiana).

Figure 8.10 provides capacitance-change correction factor formulas for a quadratical cross-section tank configuration.

For top-mounted probes near and parallel to a wall

$$\Delta C_q = 0.96 \, \Delta C_c \tag{8.7}$$

where ΔC_q is the change in capacitance for the quadratical system. And for the probe situated near a wall and $s \gg d$ (where s is the probe distance from the wall and d is the probe diameter)

$$\Delta C = \frac{(\epsilon_r - 1)}{4.15 \, \lg \frac{4s}{d}} \tag{8.8}$$

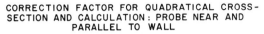

CORRECTION FACTOR FOR QUADRATICAL CROSS-
SECTION AND CALCULATION : PROBE NEAR AND
PARALLEL TO WALL

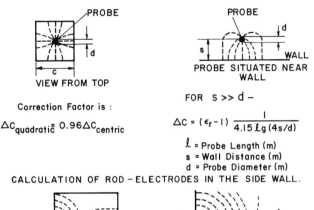

VIEW FROM TOP

Correction Factor is :

$$\Delta C_{quadratic} \equiv 0.96 \Delta C_{centric}$$

PROBE SITUATED NEAR
WALL

FOR s >> d –

$$\Delta C = (\epsilon_r - 1) \frac{l}{4.15\, lg\,(4s/d)}$$

l = Probe Length (m)
s = Wall Distance (m)
d = Probe Diameter (m)

CALCULATION OF ROD – ELECTRODES IN THE SIDE WALL.

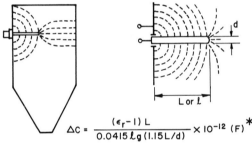

L or l

$$\Delta C = \frac{(\epsilon_r - 1)\, L}{0.0415\, lg\,(1.15L/d)} \times 10^{-12}\ (F)\ ^*$$

* d >> L , d and L in meters (m). ΔC only has to be
greater than 3 pF ; beyond that there is no influence
in ϵ changing.

Figure 8.10 Capacitance change formulas and correction factors
for quadratic cross-section container system and for electrode
(probe) installation through the side wall of the vessel. (Courtesy
of Endress & Hauser, Inc., Greenwood, Indiana).

For a probe mounted in the side wall of the vessel

$$\Delta C = \frac{(\epsilon_r - 1)\, L}{0.0477 \times 10^{-12}\ \frac{gL}{d}} \tag{8.9}$$

for d << L where L is defined in Figure 8.10.

Mode (3) This calibration procedure is followed if the dielectric constant is unknown and the tank cannot be emptied. The probe capacity in air must be estimated and the controller adjusted with the use of a decade box to the computed capacitance. While the probe is connected, the span potentiometer should be adjusted in accordance with how full the vessel is.

For conductive liquids, Mode (1) calibration is identical to Mode (1) for nonconductive liquids. For Mode (2) one must compute the probe capacity in air and adjust the controller zero, using a decade box to the computed capacitance. Once the saturation capacity of the probe is known, the change in capacitance can be computed and an external capacitor can be used for the full adjustment. Mode (3) is the same for conductive fluids as for nonconductive.

ELECTRICAL CONDUCTIVITY PROBES

Electrical conductivity can be used for high or low level detection. Conductivity-type level controls are the simplest and often the least expensive among electrical controllers.

These systems are limited to highly conductive liquids, which include most water-base materials. Oils, fats, and similar substances are generally nonconductive and therefore are not suitable applications.

A conductivity level controller generates a relatively small voltage (typically around 12V ac). One pole is connected to the partially insulated probe and the other to the container wall. The electrical resistance is measured by a Wheatstone bridge. Resistance is high (above 1 M) when the container is empty, but as soon as the conductive medium contacts the probe, it forms a low-resistance path between the probe and container wall. This change in resistance is amplified and used to operate a relay.

A common problem encountered is that the material in the vessel is often charged and the conductivity adjustment overlooked. Other important influences requiring attention are foaming, splashing, and turbulence. Most often the liquid height is of prime importance, and not the foam. There are special probes available for foaming applications. For turbulence or splashing, a still well can be employed.

Conductivity probes should never be mounted horizontally, since the liquid film that bridges the probe insulation will not run off.

Table 8.2   Materials Subject to Successful Application of Conductivity Level Controllers

Water	Cold drinks
Acids, bases, salts	Beer
Various chemicals	Wines
Sugar solutions	Beverages
Metal filings	Wood pulp, wet
Plating solutions	Soups
Milk	Sewage
Vegetable juices	Ammonia

Also, probe buildup and oil film will adversely affect operation, especially with a liquid that dries out and becomes less conductive.

Conductivity level point controllers employ the principle of electrical conductivity to sense and control the level of conductive liquids and slurries. They can also be used whenever moist or conductive solids are being handled. The controller may be used in conjunction with a single-element probe to detect either high or low levels. If two-point level control is needed, two probes are used; one for high level sensing and the other for low level sensing. The electronics may be self-contained or remotely mounted at a convenient distance from the probe.

In general, the conductivity level switch is a simple, low-cost level point controller designed for detection of conductive materials. It is provided with an adjustable sensitivity; such units can be used for detection up to 50,000 Ω. Typical substances for conductivity detection with one- or two-point level control are listed in Table 8.2.

RESISTANCE SENSORS

Resistance sensors detect material levels by the change in resistance of partially submerged electrical element. The commercial version of this technique consists of a tape-type sensor or resistor that made its first market debut about ten years ago.

The tape-type sensing element consists of a precision-wound resistor helix having from 24 to 28 contacts per foot of length. There is an outer jacket of suitable protective material which also acts as a pressure-receiving diaphragm. A breather at the top of the sensor contains chemicals for drying and for corrosion protection, and completes the isolation of the electrical system from the external environment.

The gravity pressure of the medium acts upon the jacket diaphragm and causes progressive contact of the resistive element at all points below the surface. The resistance element remains unshorted above the material surface, and it is this portion that is metered to provide the level reading.

For a typical installation, the level sensor or electric tape measure hangs from top to bottom of a tank. Two wires out of the sensor top transmit an electric resistance signal that is related to the ullage distance from the tank top down to the liquid surface. The sensor strip (lowered into place from the tank top), remains fixed in position, has no moving parts, and provides a continuous indication of the ullage resistance.

Readout instrumentation is relatively simple, since any ohmmeter can convert the sensor resistance directly to tank ullage in centimeters (with $1\,\Omega = 1$ cm, typically) [41]. Gauging systems require no deck-mounted electronics and, for small vessels, allow the use of simple battery-powered, portable ullage meters.

Figure 8.11 illustrates a typical level sensor which consists of a stainless steel base strip, insulated with plastic film on the edges and back. A flat resistance wire is wound to form a continuous helix along the full sensor length and is held away from the conducting base strip by a thin insulation layer. An outer jacket, made up of several thin plastic layers, encloses the wound resistance element and serves as the pressure-receiving diaphragm.

When the sensor is suspended in any liquid or slurry, the weight of the material compresses the jacket (refer to Figure 8.12) and shorts the helix windings below the material surface, but not above. The length of unshorted resistance helix above the material surface equals length of an ullage tape lowered from tank top to material surface, and is represented by sensor resistance R in ohms.

If, for example, the material surface goes up or down by 1 m, the length of unshorted resistance helix decreases or increases (respectively) by 1 m, and the resistance R changes by $100\,\Omega$. Resistance level sensors thus convert tank ullage distance directly into electric resistance that is relatively large in comparison to the lead-wire resistance.

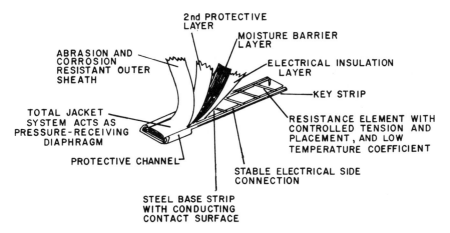

2nd PROTECTIVE LAYER

MOISTURE BARRIER LAYER

ABRASION AND CORROSION RESISTANT OUTER SHEATH

ELECTRICAL INSULATION LAYER

KEY STRIP

TOTAL JACKET SYSTEM ACTS AS PRESSURE-RECEIVING DIAPHRAGM

RESISTANCE ELEMENT WITH CONTROLLED TENSION AND PLACEMENT, AND LOW TEMPERATURE COEFFICIENT

PROTECTIVE CHANNEL

STABLE ELECTRICAL SIDE CONNECTION

STEEL BASE STRIP WITH CONDUCTING CONTACT SURFACE

Figure 8.11   Shows the composite structure of a resistance-type level sensor. (Courtesy of Metritape, Inc., Concord, Massachusetts.)

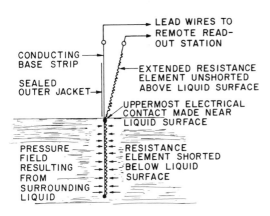

LEAD WIRES TO REMOTE READ-OUT STATION

CONDUCTING BASE STRIP

EXTENDED RESISTANCE ELEMENT UNSHORTED ABOVE LIQUID SURFACE

SEALED OUTER JACKET

UPPERMOST ELECTRICAL CONTACT MADE NEAR LIQUID SURFACE

PRESSURE FIELD RESULTING FROM SURROUNDING LIQUID

RESISTANCE ELEMENT SHORTED BELOW LIQUID SURFACE

Figure 8.12   Shows compression and progressive shorting of the submerged portion of the resistance sensor. (Courtesy of Metritape, Inc., Concord, Massachusetts.)

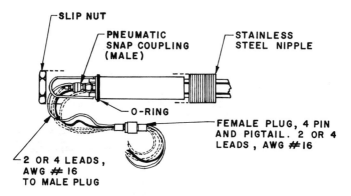

Figure 8.13   Shows typical probe configurations and design details. (Courtesy of Metritape, Inc., Concord, Massachusetts.)

Commercial resistance level gauges can also be used to measure product temperature by means of a thin, flat resistive thermal detector mounted within the level sensor. A single detector is usually placed at the sensor bottom end. It requires two additional leadwires to be brought out through the sensor top. Sometimes three detectors are spaced along the sensor back and are used to indicate temperatures at lower, middle, and upper regions of the tank. The resistive thermal detector (RTD) consists of a two-terminal wire-wound variable resistor, similar to the level sensor. It should be noted that because of their simple tank penetration and mode of operation; their ability to direct-equalize tank pressure or vacuum; and their high overload pressure tolerance, resistance sensors are well suited for service in positive-pressure, closed-tank systems.

The windings of these sensors are purely resistive; consequently, they cannot store or release transient electric energy. These level gauges thus qualify as intrinsically safe devices. They can be used in environments such as explosive gases, liquids, and dusts. Figure 8.13 shows typical probe configurations and details.

These systems have found wide acceptance in marine applications such as cargo gauging, ranging from deep supertankers to shallow barges and involving crude oil, refined petroleum products, as well as corrosive chemicals and solvents. For these applications the gauging system is principally mounted above-deck and can be refitted while the ship is in normal operating service. Pipe stills usually are needed to prevent the sensor from damage due to movement of the fluid in the ship's compartment.

There are several factors which affect system accuracy. Resistance sensors have been found to provide liquid surface indications consistently in the range of +10 mm mean deviation [41], which is consistent with the 1-cm readout resolution of most digital indicators. Teflon sensors (used for corrosive chemicals gauging) have shown inaccuracies about double the size of this figure. New refinements in Teflon jacket processing show prospects of much greater accuracies, however. The largest gauging errors are traceable to inaccurate (or elastic) tank dimensions and waves and surges on the gauged liquid surface. These factors have shown error contributions as great as 250 mm, and particularly affect the comparisons made between two independent gauging systems or between one gauging system and a reference handline measurement.

Gauge calibration can be based upon sensor length dimensions in millimeters with sensor resistance values to $0.1\,\Omega$ for sensors up to 10 m in length and to $1\,\Omega$ for those greater in length, and by use of a precision decade resistance box capable of reference resistance values to 0.01%. Such calibration accuracies are meaningful only if "as-made" tank or ship hull dimensions of equal precision can be obtained.

Further calibration difficulty results when indexing (or zero adjustment) is altered to conform to a reference handline reading. For validity, this requires precise trim measurement and correction to be made, plus careful handline measurement giving allowance for wave action, if present. Resistance gauges can be designed with any desired measure of electronic damping and will perform such averaging in a consistant and repeatable manner. This has been difficult to duplicate by means of handline gauging and attempted visual averaging.

There are a range of indicating transmitters and alarm system devices that can be incorporated with resistance sensors for level control. Figure 8.14 shows two portable indicating transmitters. Figure 8.14(A) is a single-channel indicating transmitter which provides digital readout plus an analog signal output. The unit uses a two-terminal variable resistance level (or four-terminal level/temperature) sensor to display material level or temperature. It generates a linear analog output such as 4 to 20 mA, 1 to 5 Vdc, or 0 to 4 Vdc. This is a line-powered system that can provide readings at the tank site or up to several hundred feet away. The digital display can be calibrated to make zero represent an empty tank for innage indication or a full tank for ullage indication. Level can be displayed in inches, feet, or meters accurate to 0.1%.

Figure 8.15   Digital level indicating transmitter and audible alarm sounder. The alarm unit can also be equipped with flashing light which identifies specific alarm conditions. (Courtesy of Metritape, Inc., Concord, Massachusetts.)

Figure 8.14(B) is a self-contained, battery-powered level indicating transmitter. These units can be readily added to existing tanks and provide remote readout at distances up to 1000 ft from the tank site.

Figure 8.15 shows a digital level transmitter and an audible alarm sounder. The alarm sounder can be bus-connected to one or several digital transmitters. When an alarm condition is indicated by any one of the connected transmitters, a sounder gives a loud, distinctive audible alarm.

Figure 8.14   Two types of level-indicating transmitters. (A) operates off of line voltage excitation of 115 VAC, 50-60 Hz, or 230 VAC. (B) shows a battery-operated unit. (Courtesy of Metritape, Inc., Concord, Massachusetts.)

NOMENCLATURE

A   area, in^2 or ft^2

C   capacitance, pf or f

C_a capacity of element in air, pf/ft

C_c capacity of center-mounted probe, f

C_E capacity of excentrically mounted probe, f

C_q capacity of probe mounted in a quadratical cross-section, f

D   tank diameter, ft

d   distance or probe diameter, in. or ft

g   gravitation constant, 32.2 ft/s^2

h_x level height, ft

L   length of probe end to tank wall, ft

ℓ probe length, ft

R   resistance,

Z' output impedance,

ϵ dielectric constant

ϵ_r dielectric constant of element in fluid

SUGGESTED STUDY PROBLEMS AND QUESTIONS

8.1   A parallel plate capacitor is constructed of two 1.5-in^2 plates
      separated by a distance of 0.015 in. in air. Compute the dis-
      placement sensitivity of this configuration if the dielectric con-
      stant for air is 1.0018.

8.2   Compute the capacitance and output impedance for the capaci-
      tive transducer in Problem 8.1.

8.3   A capacitance probe is mounted in a cylindrical vertical tank.
      The probe diameter is 1 in. and the tank diameter is 15 ft.
      The dielectric constant in the liquid is 9.8. Determine the
      expected changed in capacitance if the vessel is open to the
      atmosphere.

8.4   For Problem 8.3, if the overall probe length is 20 ft, what is the fluid height?

8.5   For Problem 8.3, if the probe is mounted 6 ft from the tank center line, what is the expected change in capacitance?

8.6   Explain the difference between resistance and capacitance level detection principles.

9

Ultrasonic and Sonic Instruments and Controls

INTRODUCTION

Ultrasonic sound waves are normally defined as having frequencies above 20,000 Hz (for nondestructive techniques, the frequencies are in the 1 to 15 MHz range). Basically, ultrasonics are sound waves, but are higher in tone than the frequencies normally detected by the human ear.

Ultrasonic sound waves are similar to audible sound waves in that they are mechanical vibrations involving movement of the medium in which they are travelling. In theory, any medium behaving in an elastic manner can transmit sound. Also, it has been postulated that a sound wave is propagated through a medium by particle motion. With solids, these particles do not move away from the exciting source, but rather become excited particles and excite other particles, thus producing a transmission wave. The velocity of the propagated sound wave is a function of the type of wave being transmitted and the density and elastic constants of the medium in which it is travelling.

This chapter provides an introduction to some of the properties of sound and ultrasonics along with an overview of various commercial techniques for liquid and solids level detection and control based on ultrasonics principles. Ultrasonic level methods are most often relied upon in applications which preclude the use of capacitance or conductance probes because of low dielectric properties, nonconductance, or in cases where sensors cannot be tolerated (e.g., on vessels with mixers).

PROPERTIES OF SOUND

Sound waves are a vibratory phenomenon. Conventional acoustic
measurements relate sound intensity and pressure to reference
values which correspond to the intensity and mean pressure fluctuations of the faintest audible sound at a frequency of 1000 Hz. Both
intensity and pressure levels are measured in units of decibels and
are given by the following formulae:

$$i = 10 \log_{10} \frac{I}{I_o} \tag{9.1}$$

$$P = 20 \log_{10} \frac{p}{P_o} \tag{9.2}$$

where i = sound intensity level (db),

          I = measured sound intensity (w/cm^2),

and I_o = reference sound intensity at 1000 Hz (10^{-6} w/cm^2).

In Equation (9.2) P is the sound pressure level (dB), p is the measured sound pressure (dynes/cm^2), and P_o is the reference value
(2×10^{-4} dyne/cm^2). Note that if pressure fluctuations and particle
displacements are in phase (for example, in a plane acoustic wave),
then $\lambda = \pi$, where λ is the wavelength.
    A plane acoustic wave can be described by the displacement of a
particle at a distance from the wave source and over a time period.
The following relation expresses a particle's displacement caused by
a harmonic wave.

$$\xi = \xi_o \cos \frac{2\pi}{\lambda} (ct - x) \tag{9.3}$$

where x = distance,

        ξ = particle displacement,

        ξ_o = particle displacement amplitude,

        c = velocity of sound,

        λ = wavelength,

and t = time of travel.

Equation (9.3) describes a plane wave, sometimes called a traveling wave since the particle displacement is dependent on both time and the distance from the wave source. The amplitude of the oscillatory motion will tend to decrease with distance from the wave source because of viscous dissipation of the fluid medium through which the wave travels.

Pressure fluctuations resulting from the passage of a plane sound wave are described by the following analytical expression:

$$p = \beta \frac{\delta \xi}{\delta x} = -\beta \frac{2\pi}{\lambda} \xi_0 \sin \left(\frac{2\pi}{\lambda} \right) (ct - x) \tag{9.4}$$

Equation (9.4) describes the pressure fluctuations in terms of the amplitude of the particle displacement. The parameter β is defined as the adiabatic bulk modules of the fluid.

Sound wave intensity is defined as the energy flux per unit of time per unit area. The following relations are equivalent expressions for the intensity of a plane wave:

$$I = \frac{1}{2} \rho_a c \, (\dot{\xi}_0)^2 \tag{9.5}$$

$$= \frac{1}{2} \rho_a c \omega^2 \xi_0^2 \tag{9.6}$$

$$= \rho_a c \, (\dot{\xi}_{rms})^2 \tag{9.7}$$

$$= P_{rms} \dot{\xi}_{rms} \tag{9.8}$$

$$= \frac{P_{rms}^2}{\rho_a c} \tag{9.9}$$

Where ρ_a = density of air,

ω = frequency,

ξ_0 = velocity amplitude of the wave,

$\dot{\xi}_{rms}$ = root mean square (rms) value of the particle velocity,

and P_{rms} = rms pressure fluctuation.

The intensity expressions which include pressure terms should not be applied to cylindrical or spherical waves. They can be applied to these cases as an approximation if the radius of curvature of the sound waves is large and the wave front can be approximated by a plane.

Standing waves can also be described in terms of periodic particle displacement and corresponding periodic particle velocities. If viscous dissipation effects are negligible, then, for the case of traveling waves, the amplitude of the particle displacement is the same regardless of the position x. With standing waves, the amplitude of the particle displacement follows a periodic variation with the distance x from the sound source. The term <u>standing wave</u> stems from the fact that the amplitude of the particle displacement has a periodic variation which is independent of time. The particle motion in a standing wave can be defined as follows:

$$\xi = 2\xi_0 \sin\left(\frac{2\pi x}{\lambda}\right) \cos\left(\frac{2\pi ct}{\lambda} + B\right) \tag{9.10}$$

where B is a constant.

When a sound wave moving in a medium which absorbs sound (air) strikes a "live" medium such as a wall or liquid surface, only a small portion of the sound energy penetrates the barrier and much of the wave is reflected. The reflected sound wave is, of course, an echo. If the sound wave hits an almost "dead" medium, nearly all of the sound energy is absorbed and the reflection, or return echo, is very slight. The sound absorption coefficient of a material is defined as follows:

$$\bar{\alpha} = \frac{\text{Sound energy absorbed}}{\text{Sound energy incident upon surface}} \tag{9.11}$$

The absorption coefficient is strongly dependent on frequency and is influenced by the medium properties, some of which are surface porosity, material thickness, and rigidity of the material.

One of the basic laws of optics states that

$$\text{Angle of incidence } (\alpha_E) = \text{angle of reflection } (\alpha_R) \tag{9.12}$$

This same principle also applies to sound waves. The surface of the impact material for this so-called "directed" reflection is very important to the direction of the reflected wave. A "directed"

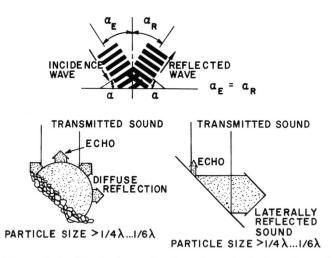

Figure 9.1   Illustrates reflective characteristics of sound waves off of different surfaces.

reflection off a sloping surface, i.e., off a steep material incline, would reflect the echo laterally and the measurement of the real round-trip time of the wave would not be accurate. Consequently, there is either a diffuse reflection or a laterally reflected sound wave off sloping surfaces, depending on the medium properties. Figure 9.1 illustrates these characteristics of sound.

Acoustic intensity (i.e., the intensity of the sound per unit area) decreases with the square of the distance ($1/r^2$) from this source, the reason being that sound is dispersed over a greater area with ever-increasing distance (see Figure 9.2). The dispersion of sound is the same for all frequencies. Figure 9.2 illustrates also that the sound pressure, which is superimposed on the normal atmospheric pressure, decreases with increasing distance ($1/r$).

Two other properties of sound worth noting before beginning discussion on level instrumentation are absorption and temperature. Absorption is the partial conversion of sound into thermal energy or heat. The degree of absorption is dependent on the propagation medium and on the frequency. At 10 kH_z, a sonic wave in a dry, dust-free environment at 20°C is attenuated by 1 to 3 dB in 50 m. An ultrasonic sound at 44 kH_z has the same degree of attenuation reached but after 1 m. Cumulative absorption is also affected by temperature, humidity, carbon dioxide, methane, and other gas

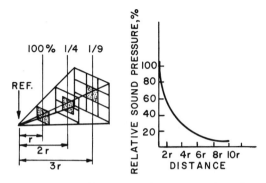

Figure 9.2 Illustrates the propagation of sound waves. Relative
sound pressure (i.e., sound pressure superimposed on the normal
atmospheric pressure) decreases with increasing distances (1/r).

concentrations in air. The decrease in sound pressure is determined
by the distance (inverse square) and by absorption.

Temperature variations can also affect sound waves. Specifically, they can impact on the sound velocity. Further discussions on
the properties of sound can be found in References 42 and 43.

ECHO-SOUNDING METHOD

A technique based on echo measurement and applied for many years
to measuring the depths of oceans and deep lakes has also found wide
application in industrial level detection applications. Figure 9.3(A)
illustrates a basic system in which a sound pulse emitter (referred
to as a transducer) is located in the bottom of a vessel filled with
liquid of which the level is to be determined. The liquid surface acts
as an acoustic reflector, and the transducer receives the reflector
of its sound pulse. The transducer is connected to a transmitter
where the sound pulse originates and to a receiver into which the
echo is fed. The transmitter and receiver are both connected with
a time interval counter which measures the elapsed time between the
emission of the sound wave and the reception of the corresponding
echo. The elapsed time can be converted into units of level of liquid
(i.e., feet, meters, etc.) which can be read from the readout instrument.

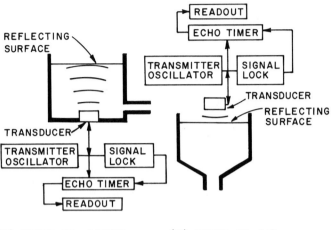

(A) SONIC IN LIQUID   (B) SONIC IN AIR

Figure 9.3   Sonic techniques used for sensing liquid or solids levels.
In (A) the sonic signal is transmitted through the liquid medium and
the liquid's surface acts as the reflecting surface. In (B) the sonic
signal travels through air to strike the liquid surface.

Figure 9.3(B) illustrates techniques that can also be used to
transmit the sonic signal through air. Here, the sensor (transducer)
is mounted over the container and emits the echo-pulses at a specified frequency (typically 10 kHz). The return time of the sonic signal is related electronically and transferred into a continuous level
indication.

Echo-based systems measure distance in terms of sound wave
time of travel. Significant erros can occur in the measurements
when temperature variations are significant. Changes in the ambient temperature can cause the sonic system to make measurement
errors in the echo round-trip time. These errors are typically on
the order of 0.18% per °C or approximately 0.1% per °F. If temperature changes are expected to be negligible, or the resulting error
acceptable, it is possible to substitute an appropriate resistor for
the middle temperature range. These systems are normally supplied with a temperature probe which compensates for the time delay
variation caused by temperature changes. In some cases, when the
level within the tank is low the error in indication caused by temperature gradients will be greater than at high levels. Consequently,
temperature compensators are strongly recommended. Temperature

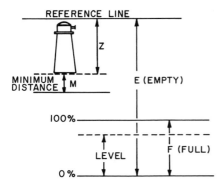

Figure 9.4   Defines major distances important to calibrating an
air-sonic echo system.

probes consist of nickel thermistors which change resistance propor-
tionally with any change in temperature.

CALIBRATION OF ECHO-SOUNDING SYSTEMS

Figure 9.4 illustrates the major distances that must be defined in
calibrating an echo-sounding system. Z represents the distance between an established reference line and the sensor cone. The reference line represents the base from which all measurements are to be
taken and is normally situated at some distance from the sensor, due
to an electronic time delay. M is the minimum distance between the
sensor cone and the 100% level. Every calibration must be conducted
with this minimum distance M or greater (note if the material reaches
a level which exceeds the 100% level, the measurement will be inaccurate). E is the distance between the reference line and the 0%
level in the tank. Distance F is the range between the 0% and 100%
level (measuring range). Finally, L represents the real level of the
material in the tank. L corresponds to the actual measuring value.
    The following relation is important to the calibration:

$$E - (Z + M) = F \qquad\qquad (9.13)$$

This relation shows that there must always be a distance between the
100% level and the sensor which is equal to or greater than M.

Since Figure 9.4 is a top-mounted transducer (i.e., sound is transmitted through air to the liquid surface), we must know something about the properties of sound in the traveling medium. The speed of sound in air as a function of temperature is given by the following formula:

$$V = 332(1 + 0.0018T) \qquad (9.14)$$

where V is the speed of sound in m/s and T is the air temperature in °C.

The time needed, then, for sound to span the entire height of the tank (or the distance to the 0% level in the tank) is

$$t = \frac{E}{V} = \frac{E}{332(1 + 0.0018T)} \qquad (9.15)$$

where t is measured in seconds.

The empty calibration is thus the calibration of the time the sound wave or pulse needs to span the maximum measuring distance and reflect back to the sensor.

This information can be translated into the electrical relay portion of the system by establishing appropriate values for resistors or potentiometer settings. A typical system would consist of an RC-section that is capable of transcribing the measured time to an appropriate electrical signal. The time formed by an RC-section would be

$$RC = t = \frac{E}{V} = \frac{E}{332(1 + 0.0018T)} \qquad (9.16)$$

or the resistance R in ohms (Ω) would be

$$R = \frac{E}{C332} \; \frac{1}{1 + 0.0018T} = \frac{R_e}{1 + 0.0018T} \qquad (9.17)$$

where R is the required resistance of the instrument for E = 1m and T = 0°C. R_e is the internal resistance of the instrument. The value of the resistance R_e may vary slightly between actual instruments, but for all commercial systems its exact value is printed on the electronic unit.

To determine the value of the resistance R for the time delay, the following is applicable:

$$R = R_e E \qquad (9.18)$$

If a resistor with a ±1% tolerance (0.5 W) is selected on the basis of its being as close as possible to the calculated value, then the difference can be calibrated by a variable resistor potentiometer. A maximum difference of ±6% may be adjustable [44].

If the chosen resistance is denoted by R_p, then by adjusting the R-potentiometer to the differential between R and R_p (in %) the proper empty calibration is obtained. To carry out the full calibration (i.e., with a full tank) it is normally sufficient to compute the ratio of distance E to F and to adjust the corresponding potentiometer to the calculated value.

As part of the system calibration, expected inaccuracies due to temperature changes should also be determined to ascertain whether temperature compensation is necessary. The full calibration inaccuracy is related to the full container. When the transducer indicates the 100% level, the distance between the unit and the liquid surface is E-F. Therefore, the distance inaccuracy is proportional to E-F. When this value is related to the full container, the resulting inaccuracy is

$$\epsilon_T \%(\text{with } \Delta T = 1°C) = e \frac{E - F}{F} \times 100 \qquad (9.19)$$

where ϵ_T is the inaccuracy (%) and e = 0.18% per °C (the effect upon sound velocity in air).

The percentage of inaccuracy ϵ_T of the actual level is, again, related to the full container:

$$\epsilon_T \%(\text{with } T = 1°C) = e \frac{E - L}{F} \times 100 \qquad (9.20)$$

Note that the empty part of the container is E-L.

Although the foregoing discussion was based on a sonic system operated in air (i.e., top-mounted transducer), similar calibration procedures are applicable to bottom-mounted systems. Some information on the properties of sound in the liquid medium is needed.

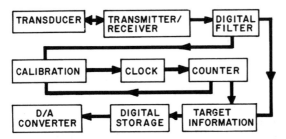

Figure 9.5 Simplified block diagram of an ultrasonic level detection system. (Courtesy of Envirotech, National Sonics, Hauppauge, New York.)

ULTRASONIC SENSOR CONFIGURATIONS AND
APPLICATIONS (LIQUID SERVICE)

As mentioned earlier, the speed of sound varies as a function of several parameters, including temperature, pressure, and density. The echo detection arrangement forms the operating principle to all ultrasonic ranging devices. Figure 9.5 provides a simplified block diagram of an ultrasonic level detection and control system. In operation, a sample of the transmitted ultrasonic signal is reflected off a "reference reflector pin" which is located at a fixed distance from the transducer face and is converted into an electrical signal upon its return to the transducer face. This sample signal is then used in a phase-lock loop circuit to calibrate a digital clock. The clock rate is calibrated to give an exact, predetermined number of pulses corresponding to the fixed distance between the transducer face and the reference reflector pin. Clock pulses are counted as the transmitted signal traverses the distance between the reference pin and the liquid surface. When the transmitted ultrasonic signal returns as an echo from the liquid surface, a counter is latched; thus, the number stored in the counter represents the true distance from the transducer (reference reflector pin) to the liquid surface. It should be noted that some of the more sophisticated systems on the market are capable of automatic and continuous compensation for temperature fluctuations.

Ultrasonic level detectors are not limited to single-sensor configurations that function on an echo recording principle. More versatile systems consist of a separate transmitter and receiver,

Figure 9.6   C-type sensor used for thick, viscous, foamy liquids or high temperature applications. (Courtesy of Envirotech, National Sonics, Hauppauge, New York.)

separated by a specified distance. In operation, the control unit generates an electrical signal that is converted to an ultrasonic signal at the transmitter transducer. When the gap is filled with liquid, this signal is transmitted across the sensor gap and reconverted to an electrical signal. The signal is amplified in the control unit and a relay is energized. When the liquid falls below the sensor gap, the signal is attenuated. With the electrical signal thus greatly reduced, the relay becomes de-energized. The relays can be used to operate indicators, event recorders, controllers, pumps, signal lights, audible alarms, valves, processing equipment, and so forth. Figure 9.6 shows one type of dual-sensor design. This particular sensor configuration works well with thick, viscous foamy liquids and for liquids at high-temperature service. Examples of other sensor configurations are illustrated in Figure 9.7.

Ultrasonic systems are relatively sophisticated. Often these systems do not require calibration. Figure 9.8 shows one manufacturer's ultrasonic liquid level switch. The unit has its sensor, control relay, and electronic components mounted in an explosionproof,

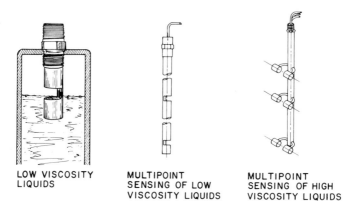

| LOW VISCOSITY LIQUIDS | MULTIPOINT SENSING OF LOW VISCOSITY LIQUIDS | MULTIPOINT SENSING OF HIGH VISCOSITY LIQUIDS |

Figure 9.7   Illustrates various ultrasonic sensor configurations for different liquid services. (Courtesy of Envirotech, National Sonics, Hauppauge, New York.)

watertight housing. This system can be used in vessels and pipes for automatic operation of alarms, pumps, or solenoid valves, and to control tank filling, emptying, or metering. It is important to note that in contrast to simple echo devices, ultrasonic systems generally are unaffected by coating buildup, clinging droplets, foam, vapor, and viscosity changes and normally are provided with automatic compensation for temperature, pressure, and other parameters,

NONPENETRATING ULTRASONIC SENSORS (LIQUID SERVICE)

Thus far we have examined single and dual ultrasonic/sonic sensor devices. All of these systems require penetration into the tank environment. There are situations, however, where it is impractical or undesirable to install a control or level sensing device in an existing installation. There are a group of nonpenetrating sensor systems that have recently appeared on the market. Figure 9.9 illustrates one of these devices.

Nonpenetrating sensors transmit an ultrasonic signal through the walls of a tube, pipe, or vessel filled with liquid but will not transmit when the vessel is filled with a gas. In operation, the control unit generates an electrical signal that is converted to an ultrasonic signal at the transmitter transducer. When the vessel is filled with

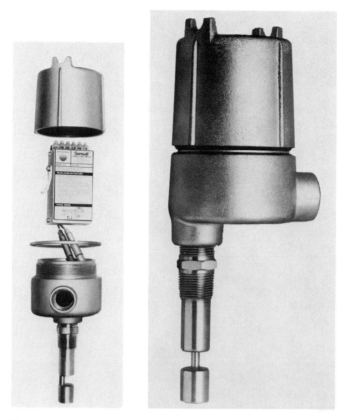

Figure 9.8   Shows two different ultrasonic liquid level switches. (Courtesy of Envirotech, National Sonics, Hauppauge, New York.)

liquid this signal is transmitted through the front walls, liquid, and rear walls to the receiver transducer and is there reconverted to an electrical signal. The signal is amplified in the control unit and a relay is energized. As with the other dual-sensor designs, when the liquid falls below the sensor gap level the signal is attenuated by the air or gas (i.e., the electrical signal is greatly reduced) and the relay becomes de-energized.

These sensors can be mounted to a vessel wall either permanently or temporarily and are well suited for a wide range of applications.

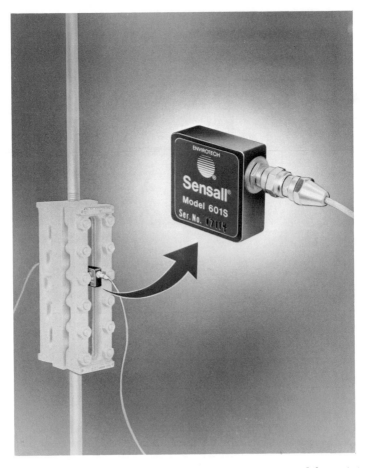

Figure 9.9   Nonpenetrating dual sensor system used for point-level control on an existing sight gauge installation. (Courtesy of Envirotech, National Sonics, Hauppague, New York.)

Figures 9.10 and 9.11 illustrate typical liquid level applications with these devices.

BIN LEVEL SENSORS

Echo detection or single-sensor units are not applicable for solid materials. Here, transmitted sound must be picked up by a receiver

DETECTS AND ACTUATES
AN ALARM WHEN ANY
LIGHT HYDROCARBON
ADDITIVE LIQUID LEVELS
DROP TO PREDETERMINED
LEVELS IN HOLDING
RESERVOIRS.

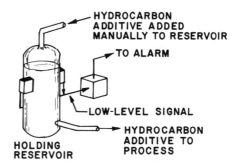

HYDROCARBON
ADDITIVE ADDED
MANUALLY TO RESERVOIR

TO ALARM

LOW-LEVEL SIGNAL

HYDROCARBON
ADDITIVE TO
PROCESS

HOLDING
RESERVOIR

MONITORS CHEMICAL
FLOW IN ALL GLASS
PIPING SYSTEM.
SENSOR DETECTS
WHEN THE LEVEL IN
THE VERTICAL
COLUMN RISES TO
PREDETERMINED LEVEL
INDICATING EXCESSIVE
FLOW PRESSURE.

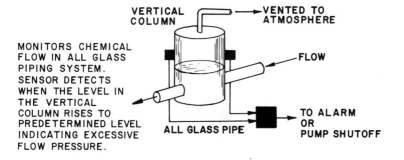

VERTICAL
COLUMN

VENTED TO
ATMOSPHERE

FLOW

TO ALARM
OR
PUMP SHUTOFF

ALL GLASS PIPE

MONITORS CRITICAL
LEVELS IN AN OIL
REFINERY DISTILLA-
TION / REACTOR
PROCESS. SENSORS
ARE MOUNTED AT
PREDETERMINED
HIGH AND LOW
LEVELS.

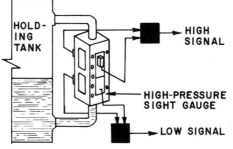

HOLD-
ING
TANK

HIGH
SIGNAL

HIGH-PRESSURE
SIGHT GAUGE

LOW SIGNAL

MONITORS AND ALARMS
WHEN THE SAND OF A FILTER
IS TOO TURBULENT DURING
BACKWASHING AND ALSO
SIGNALS WHEN THE SAND HAS
SETTLED TO INDICATE THAT
CONTINUED LAB TESTING OF
A LIQUID TOXIN CAN BE
RESTARTED. NON—
PENETRATION SENSING
ENSURES THAT NO
CONTAMINATION
OCCURS OUTSIDE
THE VESSEL.

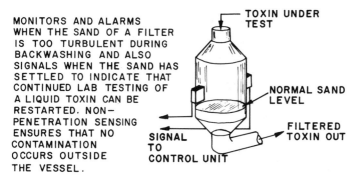

TOXIN UNDER
TEST

NORMAL SAND
LEVEL

FILTERED
TOXIN OUT

SIGNAL
TO
CONTROL UNIT

Figure 9.10   Illustrates typical applications of nonpenetrating ultrasonic level sensors.

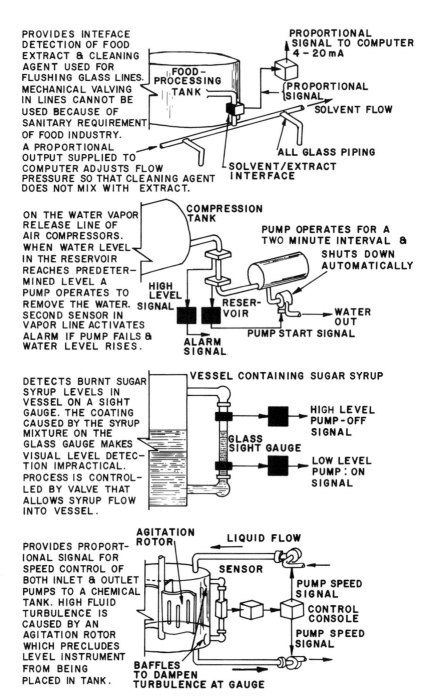

PROVIDES INTEFACE DETECTION OF FOOD EXTRACT & CLEANING AGENT USED FOR FLUSHING GLASS LINES. MECHANICAL VALVING IN LINES CANNOT BE USED BECAUSE OF SANITARY REQUIREMENT OF FOOD INDUSTRY. A PROPORTIONAL OUTPUT SUPPLIED TO COMPUTER ADJUSTS FLOW PRESSURE SO THAT CLEANING AGENT DOES NOT MIX WITH EXTRACT.

FOOD-PROCESSING TANK

PROPORTIONAL SIGNAL TO COMPUTER 4 - 20 mA

PROPORTIONAL SIGNAL

SOLVENT FLOW

ALL GLASS PIPING

SOLVENT/EXTRACT INTERFACE

ON THE WATER VAPOR RELEASE LINE OF AIR COMPRESSORS. WHEN WATER LEVEL IN THE RESERVOIR REACHES PREDETERMINED LEVEL A PUMP OPERATES TO REMOVE THE WATER. SECOND SENSOR IN VAPOR LINE ACTIVATES ALARM IF PUMP FAILS & WATER LEVEL RISES.

COMPRESSION TANK

PUMP OPERATES FOR A TWO MINUTE INTERVAL & SHUTS DOWN AUTOMATICALLY

HIGH LEVEL SIGNAL

RESERVOIR

WATER OUT

PUMP START SIGNAL

ALARM SIGNAL

DETECTS BURNT SUGAR SYRUP LEVELS IN VESSEL ON A SIGHT GAUGE. THE COATING CAUSED BY THE SYRUP MIXTURE ON THE GLASS GAUGE MAKES VISUAL LEVEL DETECTION IMPRACTICAL. PROCESS IS CONTROLLED BY VALVE THAT ALLOWS SYRUP FLOW INTO VESSEL.

VESSEL CONTAINING SUGAR SYRUP

HIGH LEVEL PUMP-OFF SIGNAL

GLASS SIGHT GAUGE

LOW LEVEL PUMP : ON SIGNAL

PROVIDES PROPORTIONAL SIGNAL FOR SPEED CONTROL OF BOTH INLET & OUTLET PUMPS TO A CHEMICAL TANK. HIGH FLUID TURBULENCE IS CAUSED BY AN AGITATION ROTOR WHICH PRECLUDES LEVEL INSTRUMENT FROM BEING PLACED IN TANK.

AGITATION ROTOR

LIQUID FLOW

SENSOR

PUMP SPEED SIGNAL

CONTROL CONSOLE

PUMP SPEED SIGNAL

BAFFLES TO DAMPEN TURBULENCE AT GAUGE

Figure 9.11   Illustrates typical applications of nonpenetrating ultrasonic level detectors.

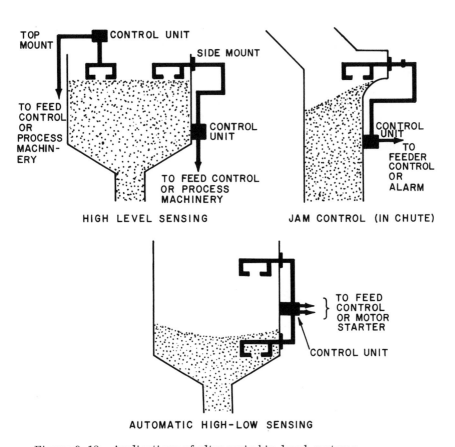

Figure 9.12   Applications of ultrasonic bin-level systems.

and fed back to the transmitter. The sensor detects the level of dry product when its ultrasonic beam is interrupted by the presence of the product at the point of installation (i.e., a large difference in signal output is provided between the presence and absence of product in the sensor gap). These systems are commercially supplied for sidewall or top mounting. Sidewall installations can be made at almost any desired point. For low-position mountings, sometimes a baffle is recommended to help divert the flow from directly contacting the sensor (this protects the sensor from surges of compact product or heavy materials). Figure 9.12 illustrates typical applications

and probe configuration. Usually for large bins or vessels handling
dry materials multipoint level detection is needed.

The advantages of using ultrasonic level detectors in bin installations include the fact that it has no moving parts (and therefore
minimal maintenance); its sensors are not affected by vibrations normally encountered in bin walls (special flanged gasket connections are
used to minimize vibrations); its systems include variable time delay
on release which prevents false alarm due to sliding, surging, or
tumbling products; its units operate at energy levels well below that
considered hazardous by any code, and these systems are applicable
to moderately high bin pressure levels (typically 50 lb/in^2) [45].

It should be noted that ultrasonic level detectors for solids applications are subject to fouling. Slight product buildup, typically
0.2 to 13 mm in thickness, depending on the type of product, can
normally be tolerated. Excessive caking can result in erroneous signals. For example, moist powders tend to cake heavily and pack in
the sensor gap. Product material properties should be considered
carefully when selecting level detection and control devices for bins.
Table 5.1 in Chapter 5 provides handling properties of various dry
materials.

NOMENCLATURE

B	constant in Equation (9.10)
C	capacitance, f
c	velocity of sound, m/s
E	length defined in Figure 9.4, m
e	error band, % [see Equation (9.19)]
F	length defined in Figure 9.4, m
I	sound intensity, w/cm^2
i	sound intensity level, dB
L	length defined in Figure 9.4, m
M	length defined in Figure 9.4, m
P	sound pressure level, dB
p	sound pressure, $dynes/cm^2$

R	resistance, Ω
R_e	internal resistance, Ω
r	distance from source, ft or m
T	temperature, °C
t	time, s
V	speed of sound, m/s
x	distance travelled, m
Z	length defined in Figure 9.4, m
$\bar{\alpha}$	sound absorption coefficient [see Equation (9.11)]
α	angle
β	adiabatic bulk modules of fluid
ϵ_T	error from temperature variations, %
λ	acoustic wavelength,
ξ	particle displacement
$\dot{\xi}$	velocity amplitude of wave
ρ	density, lb/ft^3
ω	frequency, s^{-1}

Subscripts

E	refers to incidence
o	refers to reference condition
R	refers to reflection
rms	refers to root mean square

SUGGESTED STUDY PROBLEMS AND QUESTIONS

9.1   Estimate the total sound intensity that results from the sound
      sources at 55, 60, and 75 dB.

9.2   A vessel is to be equipped with a top-mounted, echo-sonic indi-
      cator. The distance between the reference line and the vessel's
      bottom is 8.5 m and the measuring range of the tank is 7.0 m.
      The temperature of the liquid in the tank varies between 0 to
      30°C (i.e., average temperature 15°C ±15°C). How much error
      will be introduced to measurements at the 60% indication level
      if no arrangements for temperature compensation is made?
      How much at the 30% indication level?

9.3   What is the difference between traveling and standing waves?

9.4   What is ultrasonic sound?

9.5   Compute the energy flux and rms pressure variation corre-
      sponding to a sound intensity of 115 dB.

9.6   List several factors that might cause measurement inaccura-
      cies with ultrasonic level detectors.

10

Miscellaneous Level Measurement
and Control Methods

INTRODUCTION

This chapter describes four techniques for level detection: infrared
or photoelectric methods, microwave or radar systems, nuclear-
type level detectors, and thermal techniques. With the exception of
thermal techniques, the remaining three methods are generally more
expensive than many of the techniques previously described. Photoelectric methods are somewhat of an exception in that they have been
incorporated into a variety of other process control schemes; consequently, their costs have dropped in recent years. It should be noted
that there are many difficult level control problems encountered in
the industrial environment. Therefore, depending on the nature of
the problem, the more sophisticated and initially more costly instrumentation may be the only practical solution, and often the most
economical, in the long run.

INFRARED AND PHOTOELECTRIC METHODS

Infrared detectors normally used in spectrometers consist of a bis-
muth-bismuth tin thermocouple that is encased in an evacuated housing having a potassium bromide window. Other infrared detectors
are bolometers, photoconductors, and pneumatics. A bolometer is
an instrument whose electrical resistance changes with temperature
in response to the radiant energy incident to it. Photoconductors are
similar devices in that their electrical resistance changes as a function of radiant energy falling upon them. Pneumatic systems function

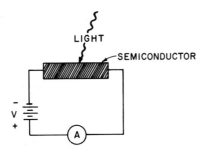

Figure 10.1   Illustrates the basic operating principle behind photo-conductive transducers.

on a gas pressure principle using a diaphragm to indicate pressure variations generated by gas volume expansion, which results from radiant energy striking the gas-pressure cell.

The electrical signal or output of an infrared detector can be recorded on a resistance-bridge, self-balancing type of strip chart recorder. The recorded output can be of two types. One type is for various wavelengths within the capability of the prism and the cell window. The other type of recording is for continuous monitoring of a flowing stream of liquid or gas at a specified wavelength. The latter is applied to continuous analysis on process flows.

Figure 10-1 illustrates the basic operating principle behind a photoconductive transducer. As shown, a voltage is impressed on a semiconductor material. When light contacts the semiconductor a decrease in resistance is experienced, causing an increase in the current as noted by the ampmeter. Photoconductive transducers, in general, have a wide range of applications. They are extremely useful for measurement of radiation at all wavelengths.

The responsivity of a detector is defined by the following

$$R_V = \frac{V_{rms}}{\gamma} \tag{10.1}$$

where V_{rms} is the root mean square (rms) output voltage and γ is the rms power incident upon the detector.

The noise-equivalent power (NEP) is the minimum radiation input that will provide a signal-to-noise ratio of 1:0. The detectivity (D) of the detector is defined as

$$D = \frac{R_V}{\emptyset} \tag{10.2}$$

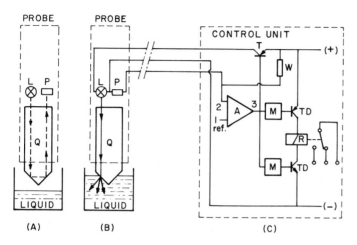

Figure 10.2   Illustrates the operating principle behind a commercial infrared, optical liquid level detector.   (Courtesy of Enraf-Nonius Service Corp., Bohemia, New York.)

where $\emptyset$ is the rms noise voltage output of the cell.   Basically, the detectivity is the reciprical of NEP.   It is often useful to express a normalized detectivity:

$$D^* = D\sqrt{A \ \Delta f} \qquad\qquad (10.3)$$

where A is the area of the detector and Δf is a noise-equivalent bandwidth. D^* provides a quantitative description of the performance of detectors so that the specific surface area and bandwidth will not affect the result.   The units of D^* are $(cm)(Hz)^{1/2}(w)^{-1}$.   The reader should refer to References 46-49 for a more in-depth treatment of this subject.

Application of infrared detection techniques to liquid level measurement and control is a fairly recent approach adopted by industry. Figure 10.2 illustrates the operating principle behind one commercial unit.   The level detection is based on the change in refraction when the conical tip of a quartz light conductor is immersed in the liquid.   Infrared light from a light-emitting diode (L) passes through the light conductor (Q) and is reflected by its conical tip if surrounded by air, gas, or vapor [Figure 10.2(A)].   The reflected light is detected by a phototransistor (P).

When the light conductor is immersed in the liquid [Figure 10.2(B)], the refraction at the tip changes and the light is dispersed in the liquid. The phototransistor (P) then receives no more reflected light.

Part (C) of this figure shows the detection circuit, which is self-checking. When the circuit is switched on, the input voltage of amplifier (A) rises via resistor W.

The output voltage (A3) rises and makes transistor (T) conducting. The light-emitting diode (L) starts emitting light. When the light is reflected, the phototransistor (P) starts conducting and reduces the input voltage (A2) to the amplifier. The amplifier output voltage (A3) drops, cutting off transistor (T). The light-emitting diode (L) is then extinguished.

The phototransistor (P) receives no more light and its resistance increases, the input voltage (A2) rises, and the light reaches the phototransistor (P) again.

Under these conditions, when the conical tip of the sensor is not immersed, the entire system will cycle. The pulsating output of the amplifier triggers two monostable multivibrators (M).

Each multivibrator controls independently a relay driver (TD). As long as the multivibrators (M) are regularly triggered, the transistor (TD) remains conductive and keeps the control relay (R) energized.

The control relay has a voltage-free output contact for external use. The cycling stops when the phototransistor receives no more light, either by immersion of the tip or by failure of the supply or any component in the system.

Figure 10.3 illustrates the main features of the probe described. This type of unit is designed for use with low-pressure storage tanks, processing vessels, pipelines, cargo tanks with crude oil, chemicals, liquified gases, ballast tanks, fuel oil tanks, double-bottom tanks, etc. The probe can be installed at the top of the vessel (maximum angle, 20° from vertical) or at the side of the tank (maximum angle, 20° from horizontal) [50]. It can be screwed into the tank wall directly, or flanged.

Infrared level detection systems are capable of operating in almost all liquids. Measurements are not affected by liquid viscosity, density, conductivity, dielectrics, or color. These systems are also intrinsically safe for use in flammable liquids. The accuracy of these systems are reported to be relatively high (typically ±1 mm) [50].

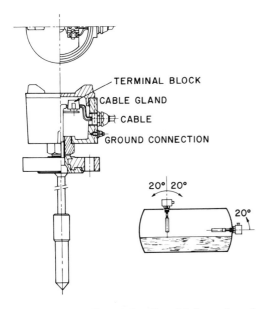

Figure 10.3   Shows details of infrared probe and mounting arrangement. (Courtesy of Enraf-Nonius Service Corp., Bohemia, New
York.)

Photoelectric control systems in general have been extended to
a wide variety of industrial applications in recent years. All photoelectric controls consist of four parts; namely, a light source, a
photosensor, an amplifier, and an output device such as a relay.
The most common type of light source is a miniature incandes-
cent bulb. These are available as either 10,000-h replaceable lamps
or 40,000-h (or more) nonreplaceable lamps. When longer life, vi-
bration, or the need for infrared light exists, then a solid-state light
source in the form of a light-emitting diode (LED) is widely used.
When ambient light is a problem, then a pulsed or modulated LED is
used. The most widely used photosensors are cadmium selenide
photocells and phototransistors. The photocells have higher sensitivity, or gain, than the phototransistors but the phototransistors
have much higher response time—generally in the order of a few
microseconds.

Figure 10.4 illustrates a few industrial applications. The light sources and photosensors may be arranged in a number of different ways. The most common arrangement is the interrupted light system in which the light source and photosensor are mounted on opposite sides of the object to be detected and the object breaks the light beam as it passes through [Figure 10.4(A)]. The reflected light system consists of a light source and photosensor mounted on the same side as the object to be detected [(B) and (C)]. The retroreflective light system is distinguished by the light source and photosensor mounted coaxially; by means of a special retroreflective target, the light beam returns on the same path from which it arrived. [Figure 10.4(D)].

A variety of amplifiers with different functions and modes of operation are on the market. Dark-operated amplifiers energize their output when the light beam is broken and fails to strike the photosensor. Light-operated amplifiers energize their output when the light beam strikes the photosensor. On-off amplifiers energize their output immediately when the light beam is broken or restored. Time-delay amplifiers (delay on, delay off, and monopulse) energize their output after a predetermined time following the breaking or restoration of the light beam. In the delay-on amplifier, the output is energized only after a predetermined time following the breaking or restoration of the light beam. In the delay-off amplifier, the output is energized immediately when the light beam is broken or restored and remains energized for a preset period of time after the input signal is removed. In a monopulse amplifier the output is energized immediately when the light beam is broken or restored and remains energized for a preset period of time, regardless of the length of time the input is activated. A variety of double delays is available combining the three basic delays described.

There are an assortment of outputs available to meet the varied system requirements. A few include electromagnetic relays, solid-state relays (triacs) for ac loads, solid state relays (power transistors) for dc loads, voltage outputs to drive an inductive load such as a relay or solenoid, and an open collector current sink output to interface with the user's electronics. Generally, the use of an electromagnetic relay as an output requires that a signal duration of 15 ms be present in order to energize the relay.

MICROWAVE (RADAR) LEVEL CONTROL SYSTEMS

Microwave controls are noncontact sensing systems. The transmitter (source) consists basically of a power supply, pulse modulator,

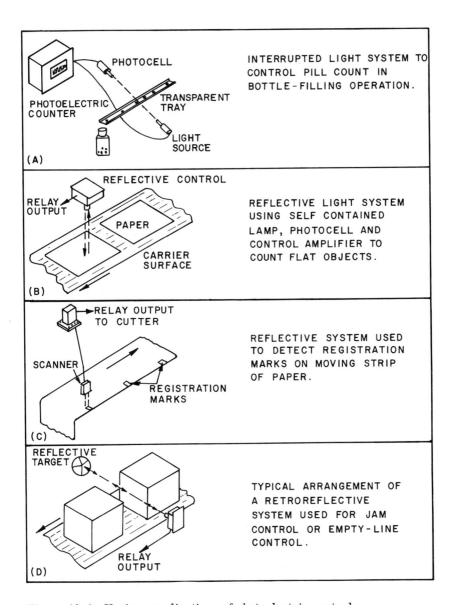

Figure 10.4   Various applications of photoelectric controls.

oscillator, and directional antenna. The receiver consists of a directional antenna, a microwave mixer cavity with a barrier diode detector, a high gain, low noise amplifier, a pulse coding network, a voltage comparator circuit, and a relay driver circuit.

Delevan Corporation [51] manufactures a system which operates as follows: in the transmitter, line voltage is converted to a well-regulated and -filtered 12V dc supply. It is then pulsed randomly at about 1 kHz by the pulse modulator circuit. This circuit is included to permit pulse discrimination circuitry to be used. In addition, pulsing at a 10% duty cycle safely permits peak transmitted power levels 10 times greater than permitted under continuous wave operation. The pulsed dc is fed to a Gunn oscillator in the antenna assembly, where the 12V dc, 1-kHz square wave is converted to a pulsed X-band (10.525 GHz) microwave signal. This signal is radiated by the directional antenna, which is typically a 10-dB gain horn with a beam spread of approximately 40%.

In the receiver, the signal is received by a directional antenna and coupled to a mixer cavity containing a detector diode. The diode converts the low level microwave signal to a low level pulsed dc, which is then amplified by an adjustable gain, low noise ac amplifier to a 0-6V dc control signal. This system is interconnected and uses pulse discrimination coding. In these systems the receiver is on only when the transmitter is on, thus the system is immune to false triggering from stray microwave signals from adjacent transmitters or other random sources of microwave interference. The level of the amplified received signal (0-6V dc) is compared with a preset value in a voltage comparator circuit. When the signal level received exceeds the comparator set point, an output signal is initiated which it processes through time delay circuits to drive the output relay.

Product materials encountered in industry have varying effects on microwave signals. For example, low level microwaves cannot penetrate metals but are reflected by them. Microwaves are absorbed almost entirely by water, and to varying degrees, by water-based solutions or products that have a significant moisture content such as grain, wood products, etc.

Transmission losses decrease with increasing dielectric constants and increase with increasing conductivity. For example, air (dielectric constant of 7, conductivity of zero) transmits microwaves with no transmission loss, while sea water (dielectric constant of 55 at X-band, conductivity of 4 mhos/m) provides extreme attenuation of the microwave energy. It is the material's dielectric constant and conductivity that determine whether or not the material is a good candidate for microwave control [51].

Microwave devices are being used to control liquids and solids in tanks, bins, hoppers, and chutes. Nonconductive fiberglass tanks represent minimal losses to X-band microwaves. Sensors are mounted on the outside of the tank, opposite one another. Losses through the tank walls and from air or vapors present above the product are low. When the product level reaches the control position, the signal is attenuated significantly, causing the relay to change state. Metal tanks or hoppers must have "windows" transparent to microwave signals. Sight glasses (3-4 in. dia.) can be used on liquid storage tanks, compatible with the pressures, temperatures, and chemical properties of the stored materials. For metal vessels storing solid materials, windows can be constructed of materials such as high-density polyethylene or other similar substances compatible with the product contained therein. A partial list of materials low in loss is given in Table 10.1, these are potential candidates for windows, as well as detected objects.

The advantages of microwave level controls include the following:

The control does not contact process materials; therefore, it cannot contaminate process material, and in turn, the process material cannot affect the control through corrosion, mechanical damage, abrasion, heat, or pressure.

The control can be installed and removed without interrupting the process.

To varying degrees, the control can penetrate process buildups on the tank wall without affecting the control's operation. For example, buildups of several inches of hydrocarbon byproducts (asphalt, cement) can be tolerated in a 20-ft-dia. tank whereas only a light coating of water-based material, such as wet latex paint, can be tolerated in a similar situation [51].

It should be noted that continuous level control systems based on radar or microwaves are rarely used in industry, due to high prices. These systems are more expensive than any of those described previously. However, recent innovations have rendered radar systems more practical. There are a few applications, such as 500-ft-deep ore bunkers or high-temperature installations, where radar has many advantages over systems discussed earlier.

Table 10.1   Potential Candidate Materials for Windows on Vessels
             Using Microwave Controls

Firebrick	Nylon
Fiberglass	Paraffin
Polyethylene	Plexiglass
Polystyrene	Glass (no lead)
PVC Powder	Teflon
Lucite	Styrofoam
Lexan	Quartz
Mica	

Microwave level controls offer another solution to those problems where a noncontact level control is indicated, but where physical limitations eliminate capacitance and other inhibitions eliminate
nuclear systems.

Vertical or horizontal installation is possible with the mounting
point fastened so that it is free from vibrations. The frequencies of
transmitter and receiver should be the same and for this reason they
are always supplied as a pair, determined by the serial number on
the units. The two units must face each other and be positioned on a
common axial line.

Slight disalignment is admissible (vertically, maximum 5% of the
distance between transmitter and receiver), but microwave power
will be lost. Since the microwave beam is polarized, the units
should not be turned on their longitudinal axis, but not by 180°.

If it is impossible to install the microwave units opposite each
other, the beams can be deflected by metal reflectors placed vertically to the mounting axis of the transmitter and receiver. The distance between the units is thus reduced by approximately 10% with
one reflector, and by approximately 20% with two reflectors. Further reflection on metal parts should, however, be avoided. The
distance between transmitter-receiver and the ground or any metal
part should amount to at least 15% of the distance between transmitter and receiver. Figure 10.5 shows the lobar propagation curve of
the microwaves. To detect small objects, the distance between the

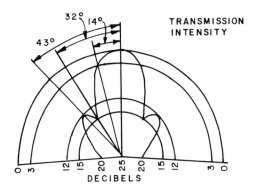

Figure 10.5   Illustrates lobar propagation of microwaves.

transmitter and receiver should be as short as possible. Materials
with a low dielectric constant cause slight attenuation and materials
with high dielectric constants produce stronger attenuation. Microwaves are best reflected on metal (refer to Table 10.2 for approximate values).

NUCLEAR-TYPE LEVEL CONTROL

Isotope source level controls are used to detect, indicate, or control
the level of almost any liquid, solid, or slurry stored in a vessel.
All elements of the system are external to the vessel so that pressure, vacuum, temperature, or materials that are highly viscous,
corrosive, abrasive, or very heavy have no influence on the system.
    Radioactive controls are initially more expensive than the other
systems described; however, their operation and maintenance costs
are so low that their implementation is often considered an economical approach. (Since there are no probes or diaphragms that require
cleaning or adjusting, there is no loss of production due to maintenance downtime.) Radioactive sensors, when properly installed,
present no hazards to operating or maintenance personnel. Many
consider them to be an unequalled solution to the most difficult level
control applications.
    The simplest nuclear system consists of a point control applica-
tion. The radioactive source in the holder emits a beam of γ rays
across the tank through the container walls to the detector. In the

Table 10.2   Attenuation of Microwaves in Various Wall Materials
(Courtesy of Endress & Hausser, Inc., Greenwood,
Indiana)

	A (thickness of wall in mm)	B (response distance in m)
PVC	30	7.5
PVC	3	10.0
Polyacryl	15	7.0
Perbunan-Rubber	20	2.0
Asbestos	2	6.8
Polyethylene	3	10.0
Glass	17	2.5
Glass	8	5.3
Wood (layers)	16	6.2
Wood (layers)	20	5.2
Wood (layers)	36	4.0

Note: A barrier is adjusted in the open air to a distance
of 10.5 m.

detector, a Geiger counter produces an electrical impulse in response to each γ photon passing through the tube. These pulses are
integrated and transformed into a dc signal proportional to the radiation received at the counter. If the level of the material is below the
location of beam, the signal received is larger than when the material is in the path of the beam (because the material will absorb or
scatter a certain amount of the γ radiation). The difference in the
two values measured is then used to control a relay system.

This relay system is essentially the same for continuous indication, the difference being that the radiation leaves the sourceholder
over an angle large enough to cover the total height range to be measured. The signal received at the detector tube varies according to

the actual level in the tank. The different path lengths are accounted for by a built-in gain compensator system.

For very large-diameter vessels, the source and detector units can be positioned across a cord. Note that extra-wide band control may require two source units and two detectors.

Common γ-ray emitting materials used as sources include radium-226 with a half-life of 1585 years, cobalt-60 (5.2 years), and cesium-137 (33 years). Radium is normally used for applications requiring 10 m or less and does not require an AEC (Atomic Energy Commission) license for its use. Cobalt and Cesium sources do, on the other hand, require AEC licensing. License application forms are generally available from the manufacturer.

It should be noted that γ rays are electromagnetic and of short wavelength and high frequency. They are measured in units of radiation intensity in air [milliroentgen (mr)]. The basis of nuclear radiation detection involves an interaction of the radiation with the detecting device, which produces an ionization reaction. The degree of ionization can be measured via appropriate electronic circuitry. The two forms of detection operations are (1) a measurement of the number of interactions of nuclear radiation with the detector and (2) measurement of the total effect of the radiation. The former is a counting process in which the energy level of the radiation is ignored. The second form of measurement may be characterized as a mean-radiation level measurement.

The reader is most likely familiar with the Geiger-Muller counter. The principle of operation behind this detector is illustrated in Figure 10.6. The anode consists of a tungsten or platinum wire, while a cylindrical tube forms the cathode for the circuit. The tube is normally filled with argon gas and a small concentration of some hydrocarbon gas. The radiation or ionization particle is transmitted through the cathode material (or through a window) and by interaction with the gas molecules; ionization of the gas results. When the voltage (V) is sufficiently high, each particle generates a voltage pulse. When the particle generates a discharge or pulse, there is a time delay before the tube can detect another particle and register another pulse. The delay consists of approximately the same amount of time needed to recharge the anode and cathode unit (i.e., to establish a new space charge in the gas).

There are a variety of other systems for detecting low level radiations. The reader can refer to References [52] and [53] for more in-depth coverage.

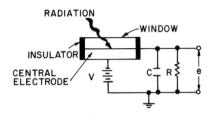

Figure 10.6   Illustrates a cylindrical-tube arrangement of a Geiger-Muller counter.

The lifetime of a detector counter tube mainly depends on the level of radiation it is exposed to. For most commercial systems, a dose rate of 0.2 mr/h gives a system lifetime in the range of 14 to 16 years. If, for example, the dose rate is increased to an average of 1 mr/h, the lifetime could be reduced to only 3 to 4 years [20]. It is important, then, to know beforehand the maximum amount of radiation likely to be used in order to extend the lifetime of the Geiger-Muller tube.

The system's response to a change in product level largely depends upon the change of the radiation on the detector created by the presence or absence of material in the beam path. Figure 10.7 illustrates various applications of nuclear-operated level sensing systems.

THERMAL TECHNIQUES

These level measuring systems are based on the differences between fluid thermal characteristics. The thermal characteristics to be measured may be either temperature or thermal conductivity.

The temperature differential technique for level measurement was at one time widely used for water level control in boiler drums. Older designs consisted of two concentric metallic tubes, each having a different thermal coefficient of expansion. The outer ends of the tube were connected to the liquid and vapor spaces of the drum. Steam in the tube was maintained at the same temperature as the fluid in the drum. The water in the tube would undergo cooling because of heat losses to the atmosphere. The relative motion of the inner tube to the outer, caused by the differences in thermal expansion, could be transmitted through a linkage to operate a feed-water

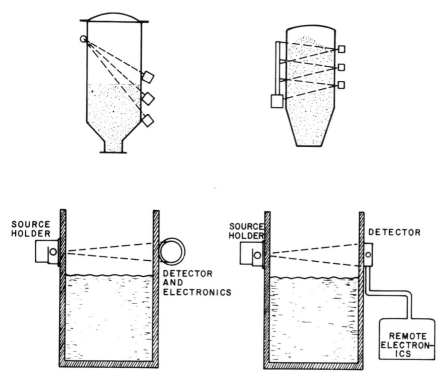

Figure 10.7   Illustrates various applications of nuclear-operated
level-sensing devices.

control valve. This design is still used today, but not as frequently.
Such systems are prone to fouling and corrosion.

Systems based on the difference in thermal conductivities between two fluids are employed as fixed-point level sensors. These
systems generally consist of an electrically heated thermistor inserted into the vessel. The thermistor's temperature and, consequently, its electrical resistance, increase as the thermal conductivity of the fluid in which it is immersed in decreases. These
devices are well suited as point level detectors for liquid-vapor interfaces, as the thermal conductivity of liquids is markedly higher
than that of vapors. Again, these systems are also prone to fouling.

11

Considerations in Process Design
and Final Control Elements

INTRODUCTION

In automatic control schemes, the level indicating signal is transmitted to a controller, whose output in turn goes to the final control elements. For many process systems the final control elements consist of valves and their driving motors. This chapter provides an overview of the various types of valves and valve actuators. The dynamic characteristics of valves and basic selection and sizing criteria are discussed also.

PROCESS SYSTEM RESPONSES TO LEVEL CHANGES

The dynamics of all fluid handling processes are strongly dependent on fluid characteristics, process vessel geometry and dimensions, and the characteristics of piping arrangement. It is necessary that these factors be given the utmost attention during the process design stages of any project, in order to minimize the overall system complexity as well as costs of regulators and control instrumentation. As explained earlier in our discussions, the flow into a holdup vessel and its discharge are related to the head Z of liquid in the vessel. A simple differential equation to describe this situation is

$$\frac{d}{dt} [AZ(t)] = Q_i(t) - Q_o(t) \qquad (11.1)$$

where Q_i and Q_o are the volumetric flow into and out of the vessel, respectively. The product $AZ(t)$ represents the volume of fluid contained in the vessel at any time t, where A is the tank's cross-sectional area.

Equation (11.1) does not describe all the possible situations, however. Liquid flow into or from a vessel can be subjected to gravitational forces, hydraulic pressures, or both. To describe these situations, hydrodynamic equations must be used. The dynamic relationship between the process parameters of pressure, hydrostatic head, and flow can be derived from the laws of conservation of mass, momentum, and energy and continuity.

Let us first examine the effects of gravity on liquid level by considering a stationary vessel of uniform cross-sectional area, discharging under its own head. If no external pressures act upon the liquid, then application of the Bernoulli equation is possible:

$$\frac{u^2}{2} + \frac{Pg}{2} + gZ = C \tag{11.2}$$

where C is a constant. Equation (11.2) describes the flow from an orifice or discharge nozzle. This can be altered to describe the flow from a discharge nozzle in a tank acting under the influence of gravitational force:

$$Q_o(t) = C_D\sqrt{2gZ(t)} \tag{11.3}$$

C_D is the nozzle's discharge coefficient and Z is equivalent to the head in the tank. The variation in liquid head as a function of the influent flow rate, Q_i, can be obtained by combining Equations (11.1) and (11.3).

$$\frac{d}{dt}[AZ(t)] = Q_i(t) - C_D\sqrt{2gZ(t)} \tag{11.4}$$

Campbell [54] notes that gravitational effects described by Equation (11.4) provide an inherent negative-feedback manipulation of the fluid level. The reason for this self-regulating effect is that Equation (11.4) states that as the influent rate $Q_i(t)$ momentarily increases the head $Z(t)$ also increases, resulting in an increase in the outflow $Q_o(t)$. This self-regulating response is nonlinear (because of $\sqrt{Z(t)}$).

Pressure can either aid or retard the flow from a tank. Pressure must be considered simultaneously with gravitational effects. The following expression describes this situation:

$$\frac{d}{dt}[AZ(t)] = Q_i(t) - C_D'\sqrt{2gZ(t) - \frac{2P(t)}{\rho}g}$$ (11.5)

The term $P(t)$ is the instantaneous pressure and is a disturbance to the system. $P(t)$ constitutes an external driving force to the system. Equation (11.5) can be solved to provide information on the liquid head as a function of time. As noted in Chapter 1, the equation can be solved to give an exact solution or can be linearized to provide an approximate solution.

Campbell [54] and Johnson [55] provide the details to the exact solution of the nonlinear response. The exact solution for a tank emptying against a back pressure as described by Equation (11.5) is

$$-\frac{t}{A} = \frac{2}{k}\left(\sqrt{Z} - \frac{Q_{io}}{k}\right)$$

$$+ \frac{2(Q_{io} + \Delta Q_i)}{k^2} \log\left(\frac{Q_{io} + \Delta Q_i - k\sqrt{Z}}{\Delta Q_i}\right)$$ (11.6)

in which $k = C_D\sqrt{2g}$ and the disturbing flow Q_i is considered as an average inflow plus a fluctuating increment $\Delta Q_i(t)$ (See Chapter 1).

To obtain a linearized solution, follow the outline given in Chapter 1 and assume that the differential form of Equation (11.4) can be expressed, using $Z = Z_0 + \delta$; where Z_0 is the average head in the tank and δ is a small change in the head (caused by flow disturbances). The general expression is

$$A\frac{dZ}{dt} + k\sqrt{Z} = \overline{Q}_i + \Delta Q_i(t)$$ (11.7)

where (-) signifies an average value. Here the disturbing inflow Q_i is considered to consist of an average inflow $\overline{Q}_i$ plus a fluctuating increment. With this assumption and for $\delta/Z_0 \ll 1$, Equation (11.7) becomes

$$A\frac{d\delta}{dt} + \frac{k\delta}{2\sqrt{Z_0}} = \Delta Q_i(t)$$ (11.8)

The back pressure P is assumed to be neglibible. Equation (11.8) is the linearized response $\delta(t)$ of a tank having an average liquid head Z_0. Consequently, if the outflow, Q_0 is under the influence of its own head, the solution must take the form of an exponential type of response for small changes in the disturbing inflow. This exponential response will have the form of $(1 - e^{-t/\theta})$, where θ is the system's time constant.

The transfer function relating the liquid head variations to ΔQ_i is as follows:

$$G = \frac{\delta(s)}{\Delta Q_i(s)} = \frac{\left(\dfrac{2\sqrt{Z_o}}{k}\right)}{\left(\dfrac{2A\sqrt{Z_o}}{k}\right)S + 1} = \frac{K}{\theta s + 1} \tag{11.9}$$

where K is defined as

$$K = \frac{2\sqrt{Z_o}}{k} = \frac{\theta}{A} \tag{11.10}$$

and k is defined as before.

The time constant θ is given by the following

$$\theta = \frac{2A\sqrt{Z_o}}{k} = \frac{2A\sqrt{Z_o}}{C_D\sqrt{2g}} \tag{11.11}$$

The transfer function G [Equation (11.9)] describes the dynamic response of the process. The time constant θ is an indication of the speed at which the tank will adjust its head after the introduction of an inflow disturbance.

From the above expressions, it should be understood that variations in liquid level are strongly dependent upon the discharge restriction to flow and tank dimensions. To illustrate this point further, Figure 11.1 and Table 11.1 give examples of several common vessel flow configurations. Figure 11.1 gives three examples of throughput configurations to holdup vessels. Included in the figure are the corresponding control block diagrams for each case. A description of the dynamic response, differential flow equations, and expressions for the transfer functions for each case are given in Table 11.1. The reader can refer to the following references for a more detailed discussion of dynamic level response and for further examples [55-58].

Table 11.1   Process Dynamics Analysis of Throughput Configura-
              tions Given in Figure 11.1

Flow Configuration	Description
Tank discharge regulated by pump speed.	Discharge flow Q_o is pumped in proportion to pump speed η. Liquid head Z varies as integral of the net flow $(\Sigma Q = Q_i - Q_o)$. Ideal integrating characteristics defined by an operator, $1/s$.
Tank discharge function of hydraulic resistance in outflow line.	Since tank's outflow limited by hydraulic resistance in outflow line, the total pressure available for driving liquid out is consumed by any pressure drop across a valve. Flow Q_o times valve resistance will consume the head Z. Inflow Q_i is system disturbance.
Tank discharge function of valve-stem position.	Resistance to outflow due to valve. Valve resistance is a function of valve-stem movement ϕ. If valve stem position is fixed, then this case reduces to the above. Difference occurs when inflow acts as a disturbance and a time-varying manipulation of the valve setting $\phi(t)$ is made. Changes in flow Q_i and changes with ϕ then control outflow head in the tank.

Flow Equation	Transfer Equation (G)	Comments
$A\dfrac{dZ(t)}{dt} = Q_i(t) - Q_o(t)$ $Q_o(t) = K_p\phi_p(t)$	$\dfrac{Z(s)}{\Sigma Q(s)} = \dfrac{1}{As}$	This type of process has no self-regulation, as outflow from vessel is not a function of the head.
$A\dfrac{dZ(t)}{dt} = Q_i(t) - Q_o(t)$ $Q_o(t) = \beta\sqrt{2gZ(t)}$ $= K'\sqrt{Z(t)}$ $\simeq \dfrac{K'}{2\sqrt{Z_o}} Z(t)$ $\dfrac{\delta Q_o}{\delta Z} = \dfrac{K'}{2\sqrt{Z_o(t)}} = K_1$ Z_o = average static head	$\dfrac{Z(s)}{Q_i(s)} = \dfrac{K_o}{\tau_o s + 1}$ where $K_o = \dfrac{2\sqrt{Z_o}}{K'}$, $\tau_o = AK_o$	Example of self-regulation in storage tank. Dynamic response characterized by first-order time response described in Chapters 1 and 2.
$A\dfrac{dZ(t)}{dt} = Q_i(t) - Q_o(t)$ $Q_o(t) = \beta(\phi)\sqrt{2gZ(t)}$ $\dfrac{\delta Q_o}{\delta t} = K_Z$; $\dfrac{\delta Q_o}{\delta \phi} = K_\phi$	$Z(s) = \dfrac{K_o}{\tau_o s + 1}(Q_i(s)$ $- K_\phi \phi(s))$ where $K_o = \dfrac{1}{kZ}$, $\tau_o = \dfrac{A}{K_\phi} = AK_o$	Inflow Q_i constitutes a disturbance, in which the valve is used to counteract the distrubance.

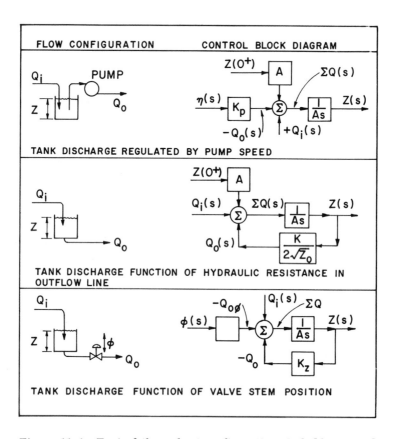

Figure 11.1 Typical throughput configurations in holdup vessels and their corresponding control-block diagrams. Refer to Table 11.1 for discussion and dynamic equations.

RELATIONSHIPS BETWEEN FLOW AND LEVEL

In liquid service, the level in a vessel and the flow through it are not independent parameters. Flow variations in fluid streams entering or discharging from a tank are brought about by the pressure and vacuum conditions in the vessel as well as in the liquid head. Fluid

flow variations, in turn, cause head changes in open tanks and can
cause both head and pressure changes in closed systems. It should
thus be realized that the regulation of liquid level precludes simul-
taneous flow regulation. The opposite is also true; i.e., during flow
regulation, surges in liquid level are inevitable.

Pumps, valves, and compressors represent the means by which
level and flow variables are established. Compressors and valves
produce pressures in fluids causing them to flow into or out of tanks
despite the various flow resistances that may exist in a process. In
contrast, valves manipulate fluid flow by varying resistance in the
flow path. A valve dissipates the fluid energy by varying flow resistance. Pumps and compressors supply potential energy to or remove
kinetic energy from the fluid energy system.

Proper selection of the level measuring instrumentation is but
one concern of the process or design engineer when designing a system; the right control regulator must also be specified. Level must
first be measured and converted into an appropriate signal Z_m from
the measuring device. This signal must then be compared against a
reference signal Z_{ref} and the difference signal $\epsilon = Z_{ref} - Z$ used
to drive an appropriate mechanism which can manipulate either the
inflow or outflow from the process, since this, in turn regulates
level.

Regulators which utilize pumps to control outflow, and thereby
control liquid level, have dynamic response. This is illustrated by
an example in Table 11.1. The process regulation is achieved by
continuous control of the pump speed. Naturally, pump speed cannot be changed instantaneously in response to changes in the error
signal ϵ. The pump's drive motor, for example, has a moment of
intertia and the motor control (regardless of whether it is electrical,
mechanical, or hydraulic) also has time lags. Time lags associated
with manipulators and various controls create instabilities in all
forms of regulators. For example, they can limit the maximum gain
that can be achieved in certain closed-loop continuous regulators and
they determine the amplitude of relaxation oscillation as well as the
frequency of the oscillatory response of on-off regulators.

Regulators employing valves to manipulate vessel outflow can
perform three modes of level control. First, they can manipulate
continuously the flow. They can also adjust the flow in an on-off
fashion (provided the valve can rapidly be both fully closed and
opened). Finally, regulators can manipulate valves so that they

maintain several intermediate positions between the limits of fully closed and fully open. All the modes of valve manipulation serve as mechanical mechanisms for adjusting the resistance in the flow system.

VALVE DYNAMICS

In automatic control, output signals from the controller are trans-mitted to the final control elements which, for many process sys-tems, consist of valves and their driving motors. Proper selection and specification of valves and valve actuators are an integral part of a process system design. Actuators are categorized on the basis of their power source with the principal types being pneumatically, hydraulically, and electrically operated.

Pneumatically operated valve motors or actuators represent the most commonly used type employed in the process industries. Fig-ure 11.2 illustrates a pneumatic valve motor whose operating prin-ciple is based on a diaphragm motor. The diaphragm is spring-loaded in opposition to the driving air pressure such that the valve-stem position is proportional to the air pressure. The diaphragm is often of the limp-type, fabricated of rubber fabric or some resist-ant material, and is normally supported by a backup plate. Dia-phragm motors typically have a maximum allowable valve-stem stroke (or travel) of 2 to 3 in. Valve actuators with longer strokes are normally of the double-acting piston-type; otherwise a rotary pneumatic motor, driving a rack or worm gear, can be used. Such actuators are capable of driving very large valves; for rotary mo-tors, valve strokes can be as much as 5 ft [55].

It should be noted that application of tank or system air pressure to the valve motor diaphragm may not always result in the desired valve-stem position. For example, the valve stem could become frozen in its seal or stuffing box and consequently may not move at all or move only sluggishly. This problem is overcome by the in-corporation of a servo mechanism, which supplies sufficient pres-sure via a pilot amplifier to the diaphragm to ensure that the valve stem achieves the proper position. An actuator of this type is re-ferred to as a valve positioner. These devices normally use a higher source of air pressure than the normal instrument air (typi-cally 100 psig).

The forcing signal transmitted to hydraulic actuators can be either a mechanical displacement or a pneumatic or electric signal

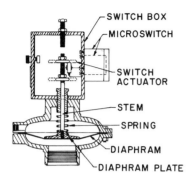

SWITCH BOX
MICROSWITCH
SWITCH
ACTUATOR
STEM
SPRING
DIAPHRAM
DIAPHRAM PLATE

Figure 11.2   Shows a pneumatic actuator. This system controls
liquid level in storage tanks and can be used to start and stop pumps,
control valves, or to set off alarms. Static pressure from the fluid
height changes the pressure supplied to the diaphragm. The pressure is opposed by spring tension. As pressure increases it causes
the diaphragm to move. Diaphragm movement raises the activator
stem. When the lower switch actuator contracts a microswitch
lever, it closes one set of contacts. These remain closed until
downward movement due to decreasing liquid level causes the upper
switch actuator to move the microswitch level in the opposite position. Switch closures may be field adjusted for the desired control
points. (Courtesy of Liquid Level Lectronics, Inc., Houston,
Texas.)

which is transformed to a mechanical displacement. The displacement in piston-type hydraulic actuators positions the cylinder valves
and admits oil to either side of the piston. This action positions the
piston shaft which then drives the valve stem.
      Another hydraulic-type actuator is the variable-delivery pump,
in which the input displacement fixes the stroke of a number of piston
pumps connected in parallel arrangement. The output-oil delivery of
the pumps is proportional to the forcing displacement and is fed to a
positive-displacement motor to produce an output displacement.
This output displacement is also proportional to the original forcing
displacement. In some cases the variable output of the piston pumps
is fed to the positive-displacement motor in such a fashion that the
rate of travel of the output displacement is proportional to the input
displacement. This results in the actuator's generating an integrat-
ing action. Hydraulic actuators are normally employed in situations
where high speeds and large forces are needed.

Electric actuators can be described as reversing motors operated by a circuit appropriate to the forcing signal. In general, these
systems are slower than pneumatic and hydraulic actuators and entirely electric control systems are often expensive.

CONTROL VALVES AND THEIR DESIGN

Control valves provide the normal mechanism for adjusting inputs of
process control systems. Proper valve selection is essential for a
well-designed level control scheme. For systems requiring relatively short time lags and having large capacities, fairly simple control systems are adequate, in that they normally permit the use of
high-sensitivity controllers with quick-opening valves. For this type
of service, flow characteristics become secondary considerations
and valve characteristics, such as size and material, become primary.

As time lag increases and capacity decreases, required control
schemes become more complex. In general, the larger the time lag,
the smaller and more precise must be the flow rate variations of the
controlling medium for given changes in process conditions. Careful consideration must therefore be given to specifying control
valves.

Several common control valve types are illustrated in Figures
11.3 through 11.7. Table 11.2 provides brief descriptions of various
control valves.

CONTROL VALVE CHARACTERISTICS

The following definitions are used to describe the operating characteristics of control valves:

Rangeability The ratio of the maximum controllable flow through
the valve to the minimum controllable flow. In practice, control
valves do not always provide complete fluid medium shutoff. Valve
seats become damaged by erosion or excessive use, or stick. Flow
in the normally closed position ranges typically from 2 to 4% of the
maximum flow (depending on valve type and size). These percentages are equivalent to rangeabilities of 50 to 25. Sliding-stem control valves have typical rangeabilities between 20 and 70.

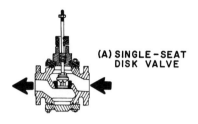

(A) SINGLE-SEAT
    DISK VALVE

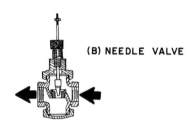

(B) NEEDLE VALVE

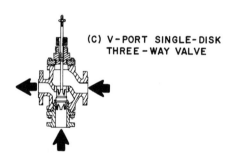

(C) V-PORT SINGLE-DISK
    THREE-WAY VALVE

Figure 11.3   Various valve bodies. (A) Single-seat disk valve is
used mainly on simple, high-sensitivity, small-time-lag applications
and for open-and-shut service or stable load conditions and constant
line pressures. (B) Needle valve is used for control of very small
flows; it is employed generally where requirements are for less than
1/2-in. valve size. (C) V-port single-disk three-way valve is suitable for mixing service.

Figure 11.4   Double-seated quick-opening inner valve with guided disks. These are used where close throttling flow is needed. Often used on low-sensitivity applications.

    Turndown   The ratio of the normal maximum flow through the valve to minimum controllable flow. A general rule applied to sizing control valves is that the maximum flow under operating conditions should provide approximately 70% of the maximum possible flow. The turndown is normally about 70% of the rangeability.

    Lift   The extent to which a valve is open. Flow through a valve depends not only on the lift but also on the pressure drop across the valve.
    The relation between flow and pressure drop for a given lift can be expressed in the form of an orifice equation for incompressible fluids.

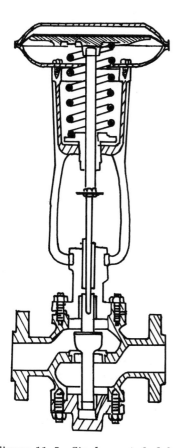

Figure 11.5   Single-seated globe-type body with solid-plug inner valve. The throttle plug controls flow rate by means of variable port openings as determined by the curvature of the plug proper.

$$Q = C \ A\sqrt{\frac{\Delta P}{\rho}} \qquad\qquad (11.12)$$

Flow control is achieved by moving the valve-stem to vary the area of flow. The gain of a valve (i.e., the change of flow for a given change in stem position) is dependent upon the change in area with stem position and also on the change in pressure drop with flow.

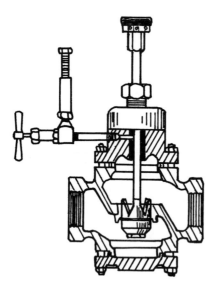

Figure 11.6   Illustrates a single-seat V-port inner valve.

Control valve manufacturers have standardized on flow coefficient [C_V in Equation (11.12)] to provide the flow capacities of valves.
The flow coefficient is generally defined as the flow of water in gallons per minute for a pressure drop of 1 lb/in across the fully open
valve. Flows for various valve stem positions are obtained from
values of C_V and the valve characteristics. These can be plotted as
percentage of maximum flow versus percentage lift or can be presented as flow chart curves as illustrated in Figures 11.8 for one
manufacturer's two-way motor valve unit. An illustrative example
on the use of this chart is also given in the figure.

The inherent valve characteristic is the relationship between the
stem position and the flow at constant pressure drop across a valve.
Linear and equal percentage are the two common characteristics encountered. The characteristic of a parabolic plug valve is referred
to as an equal-percentage characteristic. This is the kind of behavior that is desired in all characterized valves. The term <u>equal</u>
percentage refers to the fact that the flow increments resulting from
a given change in lift are the same percentage as the actual flow,
regardless of whether the valve is nearly open or nearly closed.

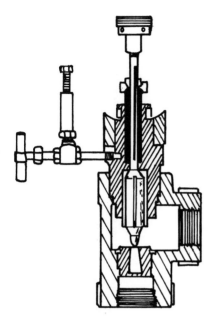

Figure 11.7   Illustrates an angle valve.

Full flow is achieved with a relatively short stem movement and consequently, the terms <u>low-lift valve</u> and <u>quick-opening valve</u> are
applied.

In sizing valves and valve capacities, the general formulas outlined below can be applied. The manufacturer should always be consulted for values of C_V and for more precise calculation procedures,
particularly when sizing valves for service in viscous liquids, flashing liquids, or gases at high pressure drops. References 59 and 60
are also good starting places for general design procedures. In
establishing the proper valve size, all flow conditions must be known
and C_V must be determined. Based on a computed C_V value, the
valve size for the type of valve under consideration can be selected
from the manufacturer's values of C_V rating versus valve size (see
Figure 11.8 for an example).

For liquid service, apply Equation (11.12) to estimate C_V and
valve capacity.

Table 11.2   Descriptions of Common Control Valve Types

Valve-Type	Description	Applications	Limitations/Disadvantages
Single-seated	Quick-opening disk-type.	Used on simple, high sensitivity, small-time-lag applications/open-and-shut service.	Impractical for throttling control if line pressures fluctuate widely, unless controller sensitivity is very high. Not recommended for throttling control above 2-in. size.
Double-seated	Upstream pressure enters body between the seats, tending to create equal upward and downward forces. Balanced valve eliminates line pressure fluctuations.	Used for close throttling control, low sensitivity applications.	Seating surfaces subject to wear. Cannot supply tight shutoff.
Throttle-plug	Double-seated, quick-opening inner valve with guided disks.	For large load changes. Handles large pressure drops well.	Limited to larger sizes.
V-port	See Figure 11.6.	Used where widely varying flow rates are encountered and where full throttling control over	

Type			
Single-seat throttling	In sizes smaller than 1 in., generally preferred to the double-seat-type. Provides tight shutoff of controlling medium.	entire flow range is needed. Has greatest controllable range. Same as single-seated.	Impractical for throttling control for lines with large pressure fluctuations.
Needle-type	See Figure 11.3/Practically any required flow characteristic is possible.	For control of very small flows.	Limited to small sizes.
Three-way-type	See Figure 11.3. Provides highly stable flows.	Service requirements include: to mix fluids (e.g., mixing hot and cold water) to feed two fluids to a vessel or pipeline; used as a throttling controller to divert fluid from one vessel or pipeline to two load demands.	

Table 11.2 (Continued)

Valve-Type	Description	Applications	Limitations/Disadvantages
Angle valves	Basically single-port valves.	Used to facilitate piping or where a self-draining piping system is needed. Used for handling fluids containing solids, slurries, and flashing fluids.	Subject to shamming.
Butterfly valves	Can be fitted with either manual or automatic valve positions. Main body may be diaphram-, cylinder-, float-, electric motor-, or solenoid-operated.	For control of low-pressure low-velocity fluids.	

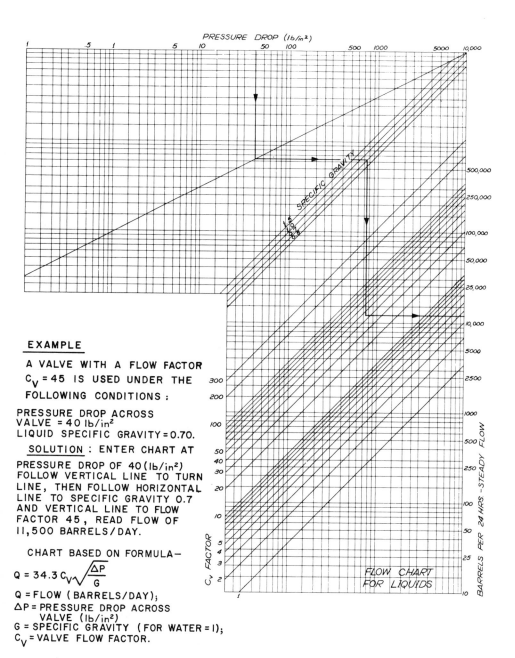

PRESSURE DROP (lb/in²)

SPECIFIC GRAVITY

BARRELS PER 24 HRS - STEADY FLOW

EXAMPLE

A VALVE WITH A FLOW FACTOR $C_v = 45$ IS USED UNDER THE FOLLOWING CONDITIONS :

PRESSURE DROP ACROSS VALVE = 40 lb/in²
LIQUID SPECIFIC GRAVITY = 0.70.

SOLUTION : ENTER CHART AT PRESSURE DROP OF 40 (lb/in²) FOLLOW VERTICAL LINE TO TURN LINE, THEN FOLLOW HORIZONTAL LINE TO SPECIFIC GRAVITY 0.7 AND VERTICAL LINE TO FLOW FACTOR 45, READ FLOW OF 11,500 BARRELS/DAY.

CHART BASED ON FORMULA—

$$Q = 34.3 \, C_v \sqrt{\frac{\Delta P}{G}}$$

C_v FACTOR

FLOW CHART FOR LIQUIDS

Q = FLOW (BARRELS/DAY);
ΔP = PRESSURE DROP ACROSS VALVE (lb/in²)
G = SPECIFIC GRAVITY (FOR WATER = 1);
C_v = VALVE FLOW FACTOR.

Figure 11.8 Liquid flow chart for a two-way motor valve system. (Courtesy of Liquid Level Lectronics, Inc., Houston, Texas.)

Table 11.3 Partial Listing of Suppliers/Manufacturers of Level-Sensing and Control Instrumentation

MANUFACTURERS	BEAM BREAKER	BUBBLER	CAPACITANCE	CONDUCTIVE	DIFFERENTIAL PRESSURE	DIAPHRAGM	DISPLACER	FLOAT	FLOAT/TAPE	PADDLE WHEEL	WEIGHT/CABLE	GAGES GLASS	GAGES MAGNETIC	MICROWAVE	RADIATION	SONIC ECHO SONAR	SONIC ECHO SONIC	ULTRASONIC	THERMAL	VIBRATION	INDUCTIVE	SWITCHES
GOULD, INC.			■																			
HARTEL CORP.		■			■																	
HARWIL CORP.							■															■
HONEYWELL PROCESS CONT. DIV.		■																				
HYDRO-TEMP. CONTROLS INC.							■															
ITT BARTON		■																				
ITT, McDONNEL & MILLER				■			■															
INTL. PRESSURE GAUGES, INC.																						
INVENTRON INDUSTRIES INC.																	■	■				
JACOBY-TARBOX CORP.												■	■									
JERGUSON GAGE & VALVE CO.												■										
JOHNSON CONTROLS, INC.																						
KENT PROCESS CONTROL, INC.			■		■									■								
KEYSTONE CARBON CO.																			■			
KING ENGINEERING CO.	■						■															
KISTLER-MORSE CORP.	Strain Gouge Measures Weight To Determine Level																					
KRATOS, INC.		■																				
LAKE SHORE CRYOTRONICS				■																		
LEEDS & NORTHRUP CO.																						
LICON DIV., ITW																						■
LIQUID LEVEL LECTRONICS INC.						■							■									
LIQUID METRONICS INC.																						
LITTON POTENTIOMETER DIV.																					■	
MADISON CO.																						
MAG-CON INC.				■																		■
MAGNETROL INTERNATIONAL							■															
MAJOR ENGINEERING CO.																						
MARINE MOISTURE CONTROL CO.	■		■												■							
MARTRON INC.		■																				
MERCOID CORP.		■					■				■											■
MERIAM INSTRUMENT DIV. OF SCOTT & FETZER																						
METRITAPE INC.	Electrical Resistance Tape																					
MICRO SWITCH (HONEYWELL)				■	Photoelectric Sensor With Fiber Optic Connector																	
MILLTRONICS INC.																						■
MONITOR MFG. INC.		■	■			■			■													
MONITOR TECHNOLOGY INC., MONITEK			■															■				
MONITROL INC.				■	Magnetostrictive																	
MONITOR MFG. CO. INC.															■						■	
MOORE INDUSTRIES INC.	Resistive Sensor																					
MOORE PRODUCTS INC.		■						■														
MURPHY, FRANK W. MFG. CO.							■															■
NATIONAL CONTROLS CORP.				■																		
NATIONAL SONICS, ENVIROTECH																		■				
NEWARK ELECTRONICS														■								
NUCLEAR RESEARCH CORP.															■							
OHMART CORP.															■							
OIL-RITE CORP.												■										■
OPCON	■																					
ORANGE RESEARCH INC.						■																
PEABODY FLOMATCHER		■																				■
PEERLESS NUCLEAR CORP.				■				■														■
PENBERTHY DIV., HOUDAILLE IND.												■										
PHOTOMATION INC.	■																					■
PNEUMERCATOR CO., INC.		■							■													

Table 11.3 (Continued)

MANUFACTURERS	BEAM BREAKER	BUBBLER	CAPACITANCE	CONDUCTIVE	DIFFERENTIAL PRESSURE	DIAPHRAGM	DISPLACER	FLOAT	FLOAT/TAPE	PADDLE/WHEEL	WEIGHT/CABLE	GLASS	MAGNETIC	MICROWAVE	RADIATION	SONAR	SONIC	ULTRASONIC	THERMAL	VIBRATION	INDUCTIVE	SWITCHES
POPE SCIENTIFIC INC.																						
PREFFERE INSTRUMENTS CO.		■							■		Remote											
PRINCO INSTRUMENTS INC.			■															■				
QUALITROL CORP.							■						■									■
RAGEN DATA SYSTEMS, INC.			■				■															■
RECHNER ELECTRONICS IND., INC.			■																■			
REVERE CORP. OF AMERICA							■															
REXNORD INSTRUMENT PRODUCTS																						
INDUSTRIAL INSTRUMENTATION				■	■													■				
ROBINSON-HALPERN CO.					■																	
ROSEMOUNT, INC.					■																	
SCANNING DEVICES INC.	■																					
SCHAEVITZ ENGINEERING							■															■
SCHNEIDER INSTRUMENT CO.					Resistive																	
SENSOTEC INC.					■																	
SETHCO DIV., MET PRO CORP.																						■
STERLING TECHNOLOGIES INC.			■															■	■			
TANK MATE CO.																						
TAYLOR INSTRUMENT CO.		■																				
TECHNICAL DEVICES CO., EXACTEL INSTRUMENT DIV.		■							■	■												
TEXAS NUCLEAR DIV., RAMSEY ENGINEERING CO.															■							■
TRI-CLOVER DIV., LADISH CO.								■														
TRIMOUNT INSTRUMENT CO.				■																		
TURCK MULTIPROX INC.			■																			■
UEHLING INSTRUMENT CO.		■			■																	
UNITED ELECTRIC CONTROLS CO.																						
VAREC DIV., EMERSON							■															
VEEDER-ROOT	■																					■
WARD INDUSTRIES INC.									■													
WARREN CONTROLS							■															
WESMAR	■															■	■					
WESTINGHOUSE ELECTRIC CORP.							■															■
ZI-TECH DIV., AIKENWOOD CORP.							■															

For gases, the following equations can be applied for valve size and valve capacity:

$$C_V = \frac{\bar{\nu}\sqrt{S_g T_a}}{1360\sqrt{(\Delta P)P_2}} \qquad (11.13)$$

$$\bar{\nu} = \frac{1360\,C_V\sqrt{(\Delta P)P_2}}{\sqrt{S_g T_a}} \qquad (11.14)$$

Figure 11.9 Suggested level instrumentation specification sheet.

where $\quad \bar{\nu}$ = flow quantity at 14.7 lb/in^2 absol. and 60°F ft/h^3,

$\quad\quad \Delta P$ = pressure drop at maximum flow lb/in^2 absol.

$\quad\quad P_2$ = the outlet pressure at maximum flow conditions lb/in^2 absol.

$\quad\quad S_g$ = specific gravity (for air = 1.0),

$\quad\quad T_a$ = flowing temperature absolute (i.e., °R = 460 + °F),

and $\quad C_V$ = valve flow coefficient.

Note that when the outlet pressure (under maximum flow, P_2) is less than half the inlet pressure at maximum flow, the term $\sqrt{(\Delta P)P_2}$ in Equations (11.13) and (11.14) can be approximated by $P_1/2$, where P_1 is the inlet pressure.
    For steam flow, the following equations can be applied:

$$C_V = \frac{W\eta_o}{3\sqrt{(\Delta P)P_2}} \tag{11.15}$$

$$W = \frac{3C_V\sqrt{(\Delta P)P_2}}{\eta_o} \tag{11.16}$$

where W is the steam mass flow (lb/h) and η_o = 1 + (0.0007 x °F superheat).
    For vapors other than steam, the following are applicable:

$$C_V = \frac{W}{63.4}\sqrt{\frac{\nu_2}{\Delta P}} \tag{11.17}$$

$$W = 63.4C_V\sqrt{\Delta P/\nu_2} \tag{11.18}$$

where W is the vapor mass flow (lb/h) and ν_2 is the specific volume at the outlet pressure P_2 lb/ft^3.
    Major considerations in specifying the proper control valves include the rangeability of the process, the maximum specific range of flows by the process, the normal range of operating loads to be controlled, available pressure drops at the valve location at both maximum and minimum flows, and the properties of the fluid being handled. References 61-68 should be consulted for further details on specifying control valves.

CLOSURE

A range of level sensing and control schemes has been discussed in
this volume. The basic principles behind their operation, along with
design and specification guidelines, have been included. No book,
however, is all-inclusive. Once a specific control scheme has been
chosen, it is best to seek assistance from the individuals who have
had the greatest experience with the mechanism under consideration.
These individuals are the manufacturers and they can supply detailed
design and operating experience to the level control problem. A
partial list of level sensing and control instrumentation manufacturers is given in Table 11.3. Also, level instrumentation forms prove
helpful in analyzing and documenting system requirements. The
Instrument Society of America has developed standard forms; however, these can be modified to suit special needs. As an example,
Figure 11.9 provides a level instrumentation specification sheet
devised by the author. A print or rough sketch of the installation to
which level control is to be applied should be included when convey-
ing information to manufacturers.

NOMENCLATURE

A	area, ft^2
C	constant in Equation (11.2)
C_D, C_D', C_V	discharge coefficient
G	transfer function
g	gravitation constant $32.2 \ \text{ft/s}^2$
K	see Equation (11.10)
k	see Equation (11.6)
P	pressure, lb/in^2
Q	volumetric flow, gal/min
s	transformed variable
s_g	specific gravity
T	temperature, °F

t	time, s
u	velocity, f/s
W	mass flow, lb/h
Z	height or level, ft
δ	increment in head, ft
ϵ	difference or error signal
η	pump speed, rpm
η_o	see Equation (11.16)
θ	time constant, s
ν	specific volume, lb/ft^3

Subscripts

i	refers to inlet condition
o	refers to outlet condition
1	refers to upstream
2	refers to downstream

Appendix

Review of Laplace Transformations

Methods of operational calculus can be applied to describing the
dynamic behavior of process systems. In particular, these mathematical tools can be used to define a system's transient and frequency responses. The classical approach to solving linear differential
equations first involves solving the homogeneous equation to obtain
a complementary function, then obtaining the forced response from
the nature of the forcing, and finally determining the constants of
integration based on the initial conditions. Laplace transformations
provide a shortcut in the solution of linear differential equations.

Laplace transformations provide a means for converting linear
differential equations into algebraic expressions. In this manner,
the simplicity of algebra can be used to solve differential equations.
Laplace transformations provide direct computation of the frequency-
response characteristics of a system from the differential equations
which define the system.

The Laplace transformation is designated by the symbol $\mathcal{L}$ and
its equation format is

$$\mathcal{L}\left[f(t)\right] = F(s) \tag{A.1}$$

Equation (A.1) states that the Laplace transform is a function of
time f(s) and that the transformation is F(s).

The inverse Laplace transform of a function of sF(s) is f(t) and
is defined as follows:

$$\mathcal{L}^{-1}\left[F(s)\right] = f(t) \tag{A.2}$$

In both equations, s is the Laplacian complex variable.

To obtain the transformation, the time-dependent function is multiplied by e^{-st} and the product is integrated over the limits from zero to infinity. That is

$$\mathcal{L}[f(t)] = \int_0^\infty f(t)e^{-st}\,dt = F(s) \tag{A.3}$$

The function $f(t)$ must be transformable, i.e., it must meet certain criteria in order to apply the transformation. The criteria for transformability are the following

1. $f(t)$ must be a real function.

2. $f(t)$ must be defined and single-valued for $t \geq 0$.

3. All discontinuities must be ordinary ones.

4. The integral given by Equation (A.3) must be convergent, i.e.,

$$\int_0^\infty f(t)\,e^{-st}\,dt < \infty$$

Laplace transforms have several properties which make them powerful mathematical tools. First they express homogeneity:

$$\mathcal{L}[Af(t)] = AF(s) \tag{A.4}$$

Laplace transforms are additive:

$$\mathcal{L}[f_1(t) \pm f_2(t)] = F_1(s) \pm F_2(s) \tag{A.5}$$

Laplace transforms express real differentiation:

$$\mathcal{L}\left[\frac{df(t)}{dt}\right] = sF(s) - f(o) \tag{A.6}$$

where $f(o)$ is the limit of $f(t)$ as t approaches o and t is always greater than or equal to o.
Similarly we can write

$$\mathcal{L}\left[\frac{d^2f(t)}{dt^2}\right] = s^2F(s) - sf(o) - \frac{df(t)}{dt}(o) \tag{A.7}$$

Laplace transforms express <u>real</u> <u>integration</u>:

$$f^{(-1)}(t) = \int f(t) \ dt = \int_0^t f(t) \ dt + f^{(-1)}(0) \qquad (A.8)$$

such that $\quad \mathcal{L}\left[\int f(t) \ dt\right] = \dfrac{F(s)}{s} + \dfrac{f^{(-1)}(0)}{s} \qquad (A.9)$

and $\quad \mathcal{L}\left[\int\int [f(t) \ dt] \ dt\right] = \dfrac{F(s)}{s^2} + \dfrac{f^{(-1)}(0)}{s^2} + \dfrac{f^{(-2)}(0)}{s} \qquad (A.10)$

Note $f^{(-2)}(0)$ is the double integral evaluated at $t = 0$. Three useful theorems that apply to Laplace transforms are

Initial value theorem

$$\lim_{s \to \infty} sF(s) = \lim_{t \to 0} f(t) \qquad (A.11)$$

Final value theorem

$$\lim_{s \to 0} sF(s) = \lim_{t \to \infty} f(t) \qquad (A.12)$$

Duhamel's theorem (or, the Convolution integral)

$$\mathcal{L}\left[\int_0^t f_1(T)f_2(1 - T) \ dT\right] = \mathcal{L}\left[\int_0^t f_2(T)f_1(t - T) \ dt\right]$$

$$= F_1(s) \ F_2(s) \qquad (A.13)$$

Finally it should be noted that if a change of scale is made in the time domain, the transform of $f(t/a)$ is a $\mathcal{L}(as)$. The value of a must be positive.

Laplace transformations allow one to convert time-dependent differential equations into algebraic equations in the variable S. The algebraic forms can be manipulated until their inversion to the time domain can be done simply. The inversion gives the solution of the original differential equation. A table of Laplace transforms such as Table 1.4 in Chapter 1 can be used to facilitate this last step.

To review, the method of solving differential equations is to transform them to the s domain [using Table 1.4 and Equation (A.3)] and then rearrange the resulting algebraic equation into the sums of terms in s which can then be inverted to the time domain by inspection (again using Table 1.4).

In process control, the Laplace transformation is best used to define the transfer function. The transfer function characterizes the dynamics of a linear system and is defined as the ratio of the transform of the response to the transform of the forcing variable. The transfer function is sometimes regarded as an operator, which by operation on the forcing signal, generates the response signal.

References

1. H. P. Kallen, <u>Handbook of Instrumentation and Controls</u>, 1st ed., McGraw-Hill, New York, 1961.

2. P. Harriott, <u>Process Control</u>, McGraw-Hill, New York, 1964.

3. R. V. Churchill, <u>Modern Operational Mathematics in Engineering</u>, 2d ed., McGraw-Hill, New York, 1958.

4. H. Chestnut, and R. W. Mayer, <u>Servomechanisms and Regulating System Design</u>, 2d ed., vol. 1, Wiley, New York, 1959.

5. F. E. Nixon, <u>Principles of Automatic Controls</u>, Prentice-Hall, Englewood Cliffs, New Jersey, 1953.

6. C. B. Moore, The inverse derivative: A new mode of automatic control, <u>Instruments</u>, 22, 1949.

7. C. L. Mamzic, Basic multiloop control systems, <u>I.S.A. J.</u>, June, 1960.

8. A. J. Young, <u>An Introduction to Process Control System Design</u>, Longmans, Green, New York, 1955.

9. J. G. Ziegler, Cascade control systems, <u>Bull. Tex. A & M Symp. Instr.</u>, Houston, 1954.

10. K. Iinoya, and R. J. Altpeter, Inverse response in process control, <u>Inc. Eng. Chem.</u>, July, 1961.

11. H. R. Wafelman, and H. Buhrmann, Maintain safer tank storage, <u>Hydrocarbon Processing</u>, January, 1977.

12. Pub. 13.22, Associated Factory Mutual Fire Insurance Companies, New York, 1970.

13. Electrical apparatus for explosive gas atmospheres, Classification of Hazardous Areas, publ. 79-10, International Electrochemical Commission, Geneva, Switzerland, 1972.

14. VbF: Verordung uber brennbare Flussigkeiten (Requirements for Flammable Liquids), Verlag Heymanns, Cologne, West Germany, 1972.

15. TRbF: Technische Regeln fur brennbare Flussigkeiten Deutscher Fachschriftenverlag (Technical Regulations for Flammable Liquids), Wiesbaden-Dot Zheim, West Germany, 1972.

16. Electrical Safety Practices, ISA Monograph 111, Instrument Society of America, Pittsburgh, Pennsylvania, 1972.

17. R.B. Bird, W.E. Stewart, and E.N. Lightfoot, Transport Phenomena, Wiley, New York, 1960.

18. J.C. Biery, Numerical and experimental study of damped oscillating manometers: I, Newtonian fluids, A.I. Ch. E.J., 9, 1963.

19. Visualized Manometry, brochure 020E:280-2, Meriam Instrument, Div. of Scott & Fetzer, Cleveland, Ohio, 1979.

20. L.E. Starratt, Level Control Instrumentation: Concepts, Theory, and Practice, Endress & Hauser, Greenwood, Indiana, 1978.

21. N.P. Cheremisinoff, and E.J. Turek, Liquid level control devices, Pollution Engineering, 7, 1975.

22. Bull. CP-3301-C, VAREC Div. Emerson Electric Co., Gardena, California, 1979.

23. E.G. Holzmann, Dynamic analysis of chemical processes, Trans. A.S.M.E., 251, 1956.

24. C.B. Schuder, The Dynamics of Level and Pressure Control, Bull. TM-7, Fisher Governor Co., 1970.

25. K. Iinoya, and R.J. Altpeter, Inverse response in process control, Ind. Eng. Chem., July, 1961.

26. Monitor Manufacturing, bull. 128, AL, Elburn, Illinois, 1979.

27. Monitor Manufacturing, bull. 122-R, Elburn, Illinois, 1979.

28. Monitor Manufacturing, bull. 124, Elburn, Illinois, 1979.

29. Chemical Engineer's Handbook, 15th ed., (R.H. Perry, and C.H. Chilton, Eds., McGraw-Hill, New York, 1973, pp. 22-45.

30. Specification sheet 801-2-E, Enraf-Nonius Delft, Bohemia, New York, October, 1976.

31. W.G. Holzbock, Instruments for Measurement and Control, 2d ed., Reinhold, New York, 1962.

32. W.J. Duncan, D/P transmitters handle variety of level measurements, Instruments and Control Systems, August 1979.

33. Computer Instruments Corp. (Hempstead, New York, Measure and control liquid level with pressure transmitters, Tech. Notes, September 1, 1976.

34. S.J. Bailey, Level sensors '76: a case of contact or non-contact, Control Engineering, July, 1976.

35. V.N. Lawford, Differential-pressure instruments: the universal measurement tools, Instr. Technol., December, 1974.

36. M.A. Georgeson, Tank depletion flow instruments, paper 76-817, Instrument Society of America, Houston, 1976.

37. J.P. Holman, Experimental Methods for Engineers, 2d ed., McGraw-Hill, New York, 1971.

38. R.J. Smith, Circuits, Devices, and Systems: A First Course in Electrical Engineering, Wiley, New York, 1971.

39. N.P. Cheremisinoff, Applied Fluid Flow Measurement: Fundamentals and Technology, Marcel Dekker, New York, 1979.

40. Level Control Probes, bull. L-7-76, Princo Instruments, Inc., Southampton, Pennsylvania, 1979.

41. A.D. Ehrenfried, Resistive Metritape Level/Temperature Gauge for Marine Closed-Tank Service, Second International Converence on Marine Transportation, Handling and, Storage of Bulk Chemicals, Monte Carlo, March 6-8, 1979.

42. L.L. Beranek, Acoustics, McGraw-Hill, New York, 1960.

43. P.N. Cheremisinoff, and P.P. Cheremisinoff, Industrial Noise Control Handbook, Ann Arbor Science, Ann Arbor, Michigan, 1977.

44. Echo-Sonic Indicator, tech. inf. bull. 6.74.12-09, Endress & Hausser, Inc., Greenwood, Indiana.

45. H.E. Soisson, Instrumentation in Industry, Wiley, New York, 1975.

46. International conference on photoconductivity, J. Phys. Chem. Sol., December 1961.

47. Reference Data for Radio Engineers, International Telephone and Telegraph Corp., New York, 1956.

48. R.J. Sweeney, Measurement Techniques in Mechanical Engineering, Wiley, New York, 1953.

49. K.S. Lion, Instrumentation in Scientific Research, McGraw-Hill, New York, 1959.

50. Specification sheet 740-E-1, Enraf-Nonius Delft, Bohemia, New York, 1979.

51. Industrial Level and Position Controls, catalog 1680-278, Delevan Electronics, Inc., Scottsdale, Arizona, 1979.

52. I. Kaplan, Nuclear Physics, 2d ed., Addison-Wesley, Massachusetts, 1963.

53. W.J. Prince, Nuclear Radiation Detection, 2d ed., McGraw-Hill, New York, 1958.

54. D.P. Campbell, Process Dynamics, Wiley, New York, 1958.

55. E.F. Johnson, Automatic Process Control, McGraw-Hill, New York, 1967.

56. G.K. Tucker, and D.M. Wills, A Simplified Technique of Control System Engineering, Minneapolis-Honeywell Regulatory Co., Philadelphia, Pennsylvania, 1958.

57. H.F. Olsen, Dynamical Analogies, Van Nostrand, New York, 1943.

58. S.E. Isakoff, Analysis of unsteady fluid flow using direct electrical analogs, Ind. Eng. Chem., March, 1955.

59. J.E. Valstar, A fresh look at selecting control valve characteristics, Control Eng., March, 1959.

60. G. L. Roth, Factors in selecting valves for compressible flow, Control Eng., December 1959.

61. M. S. Peters, and K. D. Timmerhaus, Plant Design and Economics for Chemical Engineers, McGraw-Hill, New York, 1968.

62. D. D. Perlmutter, Introduction to Chemical Process Control, Wiley, New York, 1965.

63. C. S. Beard, Control Valves, Instruments Publishing Co., Pittsburgh, Pennsylvania, 1957.

64. J. A. Wiedmann, and W. J. Rowan, Control-valve plug design, Trans. ASME, 78, 1956.

65. A. R. Aikman, Control valves: some theoretical considerations, Trans. Soc. Instr. Technol., 1, 1953.

66. R. W. Jeppson, Analysis of Flow in Pipe Networks, Ann Arbor Science, Ann Arbor, Michigan, 1976.

67. N. Andreiev, Survey and a guide to liquid and solid level sensing, Control Eng., May, 1973.

68. J. Hall, Guide to level monitoring, Instr. and Control Systems, October, 1978.

Index